THERMAL PHYS

STUDENT PHYSICS SERIES

General Editor:
Professor R.J. Blin-Stoyle, FRS
School of Mathematical and Physical Sciences
University of Sussex

Advisory Editors:
Professor E.R. Dobbs, *University of London*
Dr J. Goddard, *City of London Polytechnic*

The aim of the *Student Physics Series* is to cover the material required for a first degree course in physics in a series of concise, clear and readable texts. Each volume will cover one of the usual sections of the physics degree course and will concentrate on covering the essential features of the subject. The texts will thus provide a core course in physics that all students should be expected to acquire, and to which more advanced work can be related according to ability. By concentrating on the essentials, the texts should also allow a valuable perspective and accessibility not normally attainable through the more usual textbooks.

'At a time when many undergraduate textbooks illustrate inflation in poundage, both in weight and cost, an interesting countertrend is established by the introduction from Routledge of a series of small carefully written paperbacks devoted to key areas of physics. The enterprising authors are E.R. Dobbs (*Electricity and Magnetism*), B.P. Cowan (*Classical Mechanics*), R.E. Turner (*Relativity Physics*) and Paul Davies (*Quantum Mechanics*). The student is offered an account of a key area of physics summarised within an attractive small paperback, and the lecturer is given the opportunity to develop a lecture treatment around this core.' — Daphne Jackson and David Hurd, *New Scientist*

Already published

Quantum Mechanics, *P.C.W. Davies*
Electricity and Magnetism, *E.R. Dobbs*
Classical Mechanics, *B.P. Cowan*
Relativity Physics, *R.E. Turner*
Liquids and Solids, *M.T. Sprackling*
Electromagnetic Waves, *E.R. Dobbs*

Contents

Preface

The elegant subject of thermodynamics often arouses strong emotions amongst students: some find it easy while many find it complex, subtle and difficult. What is certainly true is that it is very easy to find the subject difficult. Too often, students are presented with a set of equations, involving partial differentials, which are memorised and then used as starting points in a derivation of a required thermodynamic relation: the TdS equations are an example in point. My experience in teaching thermodynamics over several years is that students are often at a loss as to where to start with this approach. It is as if they are placed somewhere in the middle of a thermodynamic maze with each step in the argument involving a manipulation of a partial differential being reppresented as a choice of direction in the maze. In this book, I have attempted to show how the important relations in thermodynamics can easily be obtained by starting from first principles; one need never get lost in the maze if one learns how to read the thermodynamic signposts at every junction.

Although I have written this book as a physicist, I have always been conscious of the deep understanding of thermodynamics possessed by my colleagues from other disciplines, particularly chemistry and engineering, which is brought about by their continual use of the subject. Engineers and chemists have a good 'feel' for the concepts of enthalpy and the Helmholtz and Gibbs free energies. I hope that some of this 'feel' will be communicated to the reader, particularly in Chapter 6.

The questions in Appendix 4 have been carefully selected and they do form an essential part of the book. I strongly advise the reader to try as many of them as possible. In some of them, I have added what I hope are useful hints.

There are many people I should like to thank for the ever willing help they have given me during the writing of this book. In particular, I should like to mention Norman Billingham and John Stamper of the School of Molecular Sciences at Sussex; Phil Neller of the School of Engineering; and Gabriel Barton, David Betts, Douglas Brewer, Peter Dawber, Jeremy Mcgee and Roy Turner of the School of Mathematical and Physical Sciences. I have also to thank Acorn Computers and Computer Concepts for Wordwise who helped in no small way in lessening the chore of producing the manuscript. Above all, I should like to thank the generation of students I have taught at Sussex, Oxford and the University of British Columbia, who really taught me the subject.

Introduction

The importance of thermodynamics

The science of thermodynamics was developed in the nineteenth century, mainly out of an interest in heat engines — the steam engine and the internal combustion engine. It concerns itself with the relationships between the large-scale bulk properties of a system which are measurable, such as volume, temperature, pressure, elastic moduli and specific heat. These are often called macroscopic properties. Thus thermodynamics belongs to classical physics.

Modern physics, on the other hand, attempts to explain the behaviour of matter from a microscopic or atomic viewpoint, using the techniques of quantum and statistical mechanics. You might ask, then, why we bother with this classical subject of thermodynamics.

The answer is that the modern-physics approach of quantum and statistical mechanics depends for its accuracy on the correctness of the microscopic model chosen to represent the physical system. By a microscopic model we mean a *simplified* picture of the system, consisting of a collection of small atomic-sized particles. For example a possible model of a crystal of common salt could be a collection of sodium and chlorine ions alternately placed at the corners of a stack of cubes, the forces between the ions being represented by springs. The accuracy of these models is often dubious, as must be any calculations based on them. Thermodynamics, on the other hand, is not dependent on any such microscopic model and it is important for that very reason. The results of quantum mechanics and statistical mechanics, when scaled up to macroscopic proportions, have to give results consistent with thermodynamics, and so we have an important check on our microscopic picture. However, thermodynamics by itself can give us no fine microscopic details: it can tell us only about the bulk properties of a system.

Classical thermodynamics, then, has a relevance within the framework of modern physics and is as important today as it ever was. This point was brought home by Einstein who in 1949 said:

A theory is the more impressive the greater the simplicity of its premises, the more varied the kinds of things that it relates and the more extended the area of its applicability. Therefore classical thermodynamics has made a deep impression upon me. It is the only physical theory of universal content which I am convinced, within the areas of the applicability of its basic concepts, will never be overthrown.

Chapter 1

Temperature

In this chapter we shall meet the concept of temperature, an idea which is at the very heart of thermodynamics. Indeed a definition which is often given of thermodynamics is that it is the study of the equilibrium properties of large-scale systems in which temperature is an important variable. We are all familiar, from our senses, of one body being hotter or colder than another. However, to put this concept on a sounder footing, we shall have to meet the so-called 'zeroth law of thermodynamics' which allows us to define the condition of thermal equilibrium; from this we can define temperature in an unambiguous way.

First, though, we have to define some basic concepts.

1.1 Systems, surroundings, state variables and walls

In thermodynamics we confine our attention to a particular part of the universe which we call our *system*. The rest of the universe outside our system we call the *surroundings*. The system and the surroundings are separated by a *boundary* or *wall* and they may, in general, exchange energy and matter, depending on the nature of the wall. We shall consider here the exchange of energy only, restricting ourselves to *closed systems*, i.e. where there is no matter exchange.

A very useful example of a system is a fixed mass of compressible fluid, such as a gas, contained in a cylinder with a moveable piston as shown in Fig. 1.1. We shall develop many of our ideas using this simple system as an example.

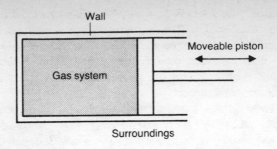

Let us first consider our system to be completely isolated from its surroundings. The degree of isolation from external influences can vary over a very wide range and it is possible to imagine walls where the isolation is complete. In practice, the rigid walls of an ordinary vacuum flask are a good approximation to completely isolating walls. It is a fact of experience that, after a time, our gas system, or any other system contained in such isolating walls, tends to an equilibrium state in which no further changes occur. In particular, the pressure P becomes uniform throughout the gas and remains constant in time, as does the volume V. We say that the gas is in the *equilibrium state (P, V)*. It is a further fact of experience that specifying these equilibrium values of the pair of independent variables P and V, together with the mass, fixes all the macroscopic or bulk properties of the gas — for example, the thermal conductivity and the viscosity. A second sample of the same amount of gas with the same equilibrium values for P and V, but not necessarily of the same shape, would have the same viscosity as the first.

These ideas can be generalised into the following definition:

An equilibrium state is one in which all the bulk physical properties of the system are uniform throughout the system and do not change with time.

We shall shortly be meeting other simple thermodynamic systems, apart from a gas, where we have to use other pairs of independent variables to specify the equilibrium state. For a stretched wire system, for example, we have to use the pair tension $\mathcal{F}$ and length L. We shall consider such other systems in Chapters 2 and 8. The important point is that we require two variables to specify the equilibrium state of a simple system and we shall call such directly measurable variables *state variables*. Other common names are *thermodynamic variables* and *thermodynamic coordinates*.

Later in this book, we shall meet some new functions of the easily measurable state variables P, V and temperature T, to be introduced in the next section. These new functions take *unique values at each equilibrium state*. Examples of such functions are the internal energy, the entropy and the enthalpy. We call such functions *state functions*. As will be shown, in an equilibrium state where they take unique values, P, V and T are functionally connected by the equation of state and so it is possible to express any one in terms of the other two. Thus these quantities are themselves state functions, but we give them the additional name of state variables because they are easily measured and enable us to specify an equilibrium state in a convenient practical way.

How then may we influence the system from outside if the walls are no longer isolating? We could bring about changes in the pressure and volume of our simple gas system in two different ways. Suppose we were to perform mechanical work on the gas system by pushing the piston in. Then the pressure and volume would in general both change, with the volume certainly changing, and we have an example of a *mechanical interaction* between the system and the surroundings. Suppose now that no mechanical interaction is allowed to occur – as would be the case if the piston were clamped, with the walls now being rigid. Consider a second cylinder, fitted with a free piston, containing the same gas with the same mass, volume and pressure as the first. Let the two cylinders be put into contact, as shown in Fig. 1.2, and let the piston of the second cylinder be pushed in. Depending on the nature of the intervening wall between the cylinders, there may or may not be changes in the pressure and volume of our gas system in the first cylinder. If there is no change, the intervening wall is said to be *adiathermal* or, more commonly, *adiabatic*: if there is a change, the wall is said to be *diathermal* and a *thermal interaction* has taken place. A wall made of metal such as copper or aluminium is a good approximation to a diathermal wall, while a good realisation of an adiabatic wall is that of a vacuum flask. Two systems in contact via a diathermal wall are said to be in *thermal contact*.

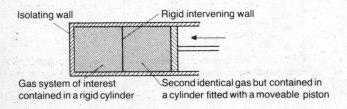

Fig. 1.2 An arrangement for determining whether or not a wall is adiabatic.

A remark should be made at this point. The reader may wonder why we do not define diathermal and adiabatic walls according to whether or not they conduct heat. The answer is that, while such walls have these properties, we cannot define them in such a way as we have not yet defined heat. This has to wait until Chapter 3.

1.2 Thermal equilibrium, the zeroth law of thermodynamics and temperature

If two thermodynamic systems such as gases are put in thermal contact, after a time no further changes in the pressures and volumes will occur, each gas being in an equilibrium state. The gases are said to be in *thermal equilibrium* with each other.

Consider the arrangement shown in Fig. 1.3, where we have three systems A, B and C, each in an equilibrium state in that the state variables have assumed constant and uniform values. Suppose now that the states of the systems are such that, when A and B

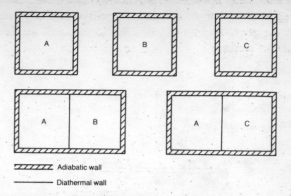

```
ZZZZZZ  Adiabatic wall
————————  Diathermal wall
```

Fig. 1.3 An illustration of the zeroth law of thermodynamics. If A and B are in thermal equilibrium upon contact, as are A and C, then so are B and C.

are brought together in thermal contact, thermal equilibrium exists in that no changes occur in the variables. Also suppose that the same is true for the systems A and C. It is an experimental observation that B and C would also be in thermal equilibrium if they were similarly brought together. This can be generalised to the statement of the zeroth law of thermodynamics:

> *If each of two systems is in thermal equilibrium with a third, they are in thermal equilibrium with one another.*

This experimental observation is the basis of our concept of temperature. It follows from the zeroth law that a whole series of systems could be found that would be in thermal equilibrium with each other were they to be put in thermal contact — a fourth system, D, which is in thermal equilibrium with system C would also be in thermal equilibrium with A and B, and so on. All these systems possess a common property which we call the *temperature, T*.

> *The temperature of a system is a property that determines whether or not that system is in thermal equilibrium with other systems.*

More formal mathematical arguments may be developed to show the existence of temperature but we shall not go into them here (see, for example, the books by Zemansky or Adkins).

1.3 Thermodynamic equilibrium

If two systems have the same temperature so that they are in thermal equilibrium, this does not necessarily mean that they are in complete or *thermodynamic equilibrium*. For this condition to hold, in addition to being in thermal equilibrium, they would also have to be in:

1. *mechanical equilibrium*, with no unbalanced forces acting; and
2. *chemical equilibrium*, with no chemical reactions occurring.

Also, any exchange of material between the systems has to be excluded for the equilibrium to be complete.

1.4 Isotherms

Let us return to our gas system contained in the cylinder with the moveable piston. Suppose that the gas is in the equilibrium state (P, V) and is in thermal equilibrium with another reference system so that the two systems have the same temperature. We can plot this state as a point on a pressure versus volume plot, which is called an *indicator diagram*.

Let the gas system be separated from the reference system. If the piston is now pushed in to take the gas to the new state (P', V') and if this new state is also in thermal equilibrium with the unchanged reference system, by the zeroth law the two states (P, V) and (P', V') are themselves in thermal equilibrium and have the same temperature. By this we really mean that two identical systems in the states (P, V) and (P', V') would be in thermal equilibrium. The locus of all such points with the same temperature is called an *isotherm*. The isotherms for an ideal gas, to be discussed in the following section, are shown in Fig. 1.4.

1.5 Equations of state

We have seen that all the bulk physical properties of a system in an equilibrium state are fixed by specifying two independent state variables, and these properties must include the temperature. For a gas this implies that there is a functional relationship

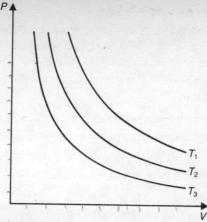

Fig 1.4 The isotherms for an ideal gas. They form a family of hyperbolae.

between P, V and T:

$$f(P, V, T) = 0$$

Such a relation is called an *equation of state*. It shows that, of the three directly measurable variables, P, V and T, only two are independent and any one may be expressed in terms of the other two. The state of the gas is equally well specified by quoting $(P, V), (P, T)$ or (V, T).

As an example of an equation of state, we may consider an ideal gas (where there are no intermolecular attractions and the molecules themselves have no volume) where the equation of state is

$$PV = nRT$$

Here n is the number of moles present and R is a constant called the universal gas constant. It follows from this equation of state that the isotherms for an ideal gas shown in Fig. 1.4 are a family of hyperbolae.

We shall meet different equations of state for systems other than a gas elsewhere in this book.

1.6 Scales of temperature

In order to give *numerical* values to different temperatures, we have to set up a method for assigning such values. Our first task is to choose a system and then to select a physical property of that system (the *thermometric variable* or *thermometric property*) that varies with temperature. A familiar example of such a choice is a mercury-in-glass thermometer, where the length of the column of mercury is the thermometric variable. In order to make the argument general, let us choose a thermometric variable which we shall call X.

Our normal choice of X is something that can easily be measured, such as the length of the column of mercury in our mercury-in-glass thermometer or the resistance of a length of platinum wire. Unfortunately, scales of temperature defined using different but familiar thermometric variables do *not* on the whole agree, although in practice the differences are small. We shall return to this point.

Let us now consider our general thermometric variable X in order to set up a scale of temperature. We call T_X the temperature on the X scale where the subscript X is to remind us that, usually, the temperature depends on the thermometric property chosen. We then define the numerical value of temperature on this scale so that the thermometric property X varies with temperature in the simplest possible way according to the linear relation

$$X = cT_X \qquad\qquad [1.1]$$

where c is a constant. The value of c is fixed by choosing an easily reproducible T_X (*a fixed point*) and assigning to it a *particular* value. The customarily chosen fixed point is the temperature at which ice, water and water vapour coexist in equilibrium; this is known as the *triple point of water*. The value given to T_X at this fixed point is 273.16 — the choice of the value 273.16 will be discussed shortly. Substituting this value for the temperature of the triple point, where the value of X is X_{TP}, in equation [1.1] gives

$$T_X = 273.16 \ (X/X_{TP}) \qquad\qquad [1.2]$$

There are two points here: (i) Equation [1.2] implies a zero of temperature on the X scale, i.e. $T_x = 0$, when $X = 0$. In practice, such a T_x may not occur if the thermometric variable does not vanish as the temperature is progressively lowered. For example, the resistance of a length of platinum wire always remains finite, tending to a constant value at the lowest attainable temperatures. The ideal gas scale, to be discussed below, does have a meaningful zero of temperature because the thermometric property used there, the pressure, eventually vanishes as the temperature is lowered.

(ii) Temperatures on the X scale are defined only in regions where equation [1.2] is meaningful. If, for example, we are using a mercury-in-glass thermometer with X being the length of the column, this equation gives a temperature only as long as there is a measurable length of mercury in the capillary. At low temperatures, if the mercury has dropped back into the bulb, equation [1.2] has no relevance. This is one reason why, in practice, the mercury-in-glass scale defined according to equation [1.2], is not used, even though in principle such a scale is possible. Instead, these thermometers are calibrated in terms of other standard ones such as those described in section 1.8.

Before 1954, temperature scales were based on the modified relation

$$X = cT_X + d \qquad [1.3]$$

where the two constants c and d had to be fixed by specifying the temperature at *two* fixed points, the steam and the ice points, which are the temperatures of boiling and freezing water at one atmosphere pressure. Since 1954, equation [1.2] has been used which requires only *one* fixed point.

It is very important to realise that different thermometers based on different thermometric variables will agree by definition only at fixed points. At other temperatures a mercury-in-glass thermometer will give slightly different values for a particular temperature than, say, a platinum resistance thermometer because the actual temperature dependencies of the two thermometric properties (the length of the mercury column and the resistance of the platinum) may be quite different. If the mercury scale and the resistance scale were truly linear, as is apparently suggested by equation [1.1], then the two scales would agree at all points.

However what equation [1.1] is saying is that the length of the mercury column increases linearly with temperature defined only according to that mercury scale. If we were to measure the temperature at which the mercury column had a given length with another type of thermometer, then there is no reason why the length of the mercury column should vary linearly with temperature measured in this second way.

Fortunately there is a class of thermometers which always agree at all points on the temperature scale – the gas thermometers. It will be shown in Chapter 4 that temperature defined according to the ideal gas scale has a fundamental significance in thermodynamics and in fact is identical to the temperature T on the absolute thermodynamic temperature scale. We shall thus call temperature, as defined on the ideal gas scale, T also. The development of all our thermodynamic relations will be in terms of T, with the understanding that it can be measured experimentally using a gas thermometer.

1.7 The gas scale

A schematic diagram of a constant volume gas thermometer is given in Fig. 1.5. The volume of the gas is kept constant by adjusting the height of the mercury column until the mercury meniscus just touches the marker at the end of the capillary tube.

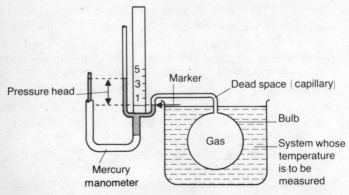

Fig. 1.5 A constant-volume gas thermometer.

The bulb of gas is immersed in a system whose temperature is to be measured and the pressure of the gas is used as the thermometric parameter. Allowance has to be made for the fact that some of the gas in the 'dead space' may be at a different temperature from that in the bulb. When this and other corrections have been made, which we shall not go into here, the gas scale temperature is determined from

$$T_{gas} = 273.16 \, (P/P_{TP}) \qquad [1.4]$$

The interesting point is that, when the amount of working gas is reduced as small as possible for measurements still to be made, *all gas thermometers give the same temperature for a given system, irrespective of the gas used.* Figure 1.6 illustrates this point for the temperature of water boiling under an external pressure of one atmosphere where the limiting value is 373.15.

We can summarise these findings by defining the ideal gas scale as

$$T = 273.16 \, \lim_{P_{TP} \to 0} \, (P/P_{TP}) \, K \qquad [1.5]$$

where we have introduced K, for kelvin as the unit of temperature on the ideal gas scale. Notice that no ° sign is written before K as °K. The connection with an ideal gas is clear when it is realised that, as the amount of working gas is reduced, the gas becomes closer to an ideal one as both the intermolecular attractions and the molecular volume are also reduced.

The apparently curious choice of 273.16 K for the temperature of the fixed point can now be understood. This value was chosen to make the size of the kelvin such that there are 100 K between the *experimentally determined* temperatures of the ice and steam points on the ideal gas scale. These temperatures have the values of 273.15 K and 373.15 K respectively. The reader should note the 0.01 K difference between the ice and triple points.

Actually, some recent measurements of the ice and steam points give values very slightly different from those just quoted, so there are not quite 100 K between these points; however, we shall ignore these *small* differences and simply note that they exist. Also, it should be remarked that the ideal gas scale can

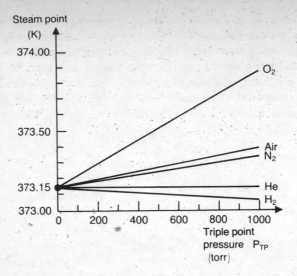

Fig. 1.6 The steam point as a function of the mass of gas used in a constant-volume gas thermometer.

equally well be defined using constant pressure thermometers rather than with constant volume thermometers. In practice, only the latter type are used and so we have confined our discussion to these.

1.8 More convenient temperature scales — the Celsius scale and the international practical temperature scale

It is convenient to have a temperature scale in which the zero is in the range of commonly encountered temperatures. The Celsius scale t is measured in °C and is related to the ideal gas scale T by

$$t(°C) = T(K) - 273.15 \qquad [1.6]$$

This means that, ignoring the small recently measured differences from the experimental values of 273.15 K and 373.15 K, discussed in the previous section, the ice point is at 0 °C and the steam point is at 100 °C. By definition, the temperature of the fixed-point triple point of water is 0.01 °C.

Because gas thermometers are cumbersome devices to use, it is also convenient to calibrate a whole series of secondary thermometers in terms of the gas scale and to use these in practice. The 1968 international practical temperature scale uses, for example: a platinum resistance thermometer between 14 K and 904 K; a thermocouple, with the two metals being platinum and an alloy of platinum with 10 per cent rhodium, between 904 K and 1340 K; and a radiation pyrometer above 1340 K.

Chapter 2
Reversible processes and work

2.1 Processes

In thermodynamics we are concerned with changes in the different state functions that occur when a system changes from one equilibrium state to another. A *process* is the mechanism of bringing about such a change. These initial and final equilibrium states are called the *end points* of the process. Pushing in the piston and compressing the gas in our cylinder from an equilibrium state (P_1, V_1) to a new equilibrium state (P_2, V_2), say, is an example of a process.

2.2 Reversible processes

There is a particular class of idealised processes which is of enormous value in thermodynamics — processes which are *reversible*. They are valuable because we can calculate changes in the state functions for any process using them. This point will be made clear later in this chapter when we consider an example of the thermodynamic method. First, let us examine what we mean by reversible processes and how they are realised.

Clearly, reversible implies that, in any such change, the system must be capable of being returned to its original state. However reversible means much more than this in that, when the system is returned to its original state, the surroundings must be unchanged too.

A clue to the conditions for reversibility can be gained by considering a pendulum being displaced from one equilibrium position to another, as in Fig. 2.1.

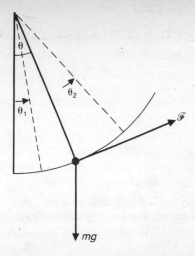

Fig. 2.1 A simple pendulum. When this is displaced it can be moved through a series of equilibrium states.

If we apply a force $\mathscr{F}$, we have to perform work in effecting this displacement. We are acting as the surroundings for the system consisting of the pendulum and the gravitational field. Suppose we pull on the bob with a force $\mathscr{F}(\theta)$, which is only infinitesimally greater than the restoring force $mg \sin \theta$ at every stage of the displacement, from θ_1 to θ_2. The pendulum then goes through a series of equilibrium states in that we could stop the process at any stage and the pendulum would stay where it is. If now we were to reduce $\mathscr{F}$ by a very small amount, the pendulum would move back the other way with the work done against us being exactly equal to the work we did in the initial displacement, and we would have reversibility. This is true provided there are no frictional forces present as there would be if the bob moved in a viscous medium. The term *quasistatic* is used for a process which is a succession of equilibrium states.

> *Reversible processes then are quasistatic processes where no dissipative forces such as friction are present.*

For a more relevant example of a reversible process, let us return to our gas-cylinder system (Fig. 1.1) and consider an isothermal reversible compression, at the temperature T, from the state (P_1, V_1) to the state (P_2, V_2). This could be achieved in the following way. The cylinder is fitted with a frictionless piston and contains the gas in the initial state (P_1, V_1). A force

$$\mathscr{F} = P_1 A$$

where A is the area of the piston, is applied to the piston to oppose the gas pressure. The walls of the cylinder are diathermal and the surroundings, at the temperature T, are so large that this temperature is unaffected by anything that we might do to the gas-cylinder system. Such surroundings are called a *thermal* or *heat reservoir*, or more briefly a *reservoir*. The external force $\mathscr{F}$ is now increased infinitesimally and the system is allowed to come to a new equilibrium state at the same temperature. This process is repeated until the final state (P_2, V_2) is reached. Because the system is always in an equilibrium state, with a well-defined P, V and T, we may plot the process on an indicator diagram as a succession of points forming a continuous curve between the end points (P_1, V_1) and (P_2, V_2) which is the isotherm at T as in Fig. 2.2. The equation of state

$$PV = nRT$$

holds for each point in the process.

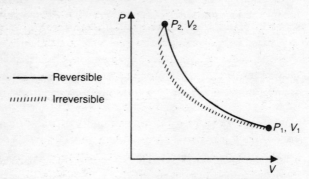

Fig. 2.2 A reversible process may be shown as a continuous line on an indicator diagram. The one shown here is an isothermal process. An irreversible process cannot be shown in this way.

In contrast, we could bring about the same change between

the same two end points by pushing in the piston violently from volume V_1 to volume V_2. There would then be turbulence, with finite temperature and pressure gradients both within the gas and between it and the surroundings. Although the gas would eventually settle down to the equilibrium state (P_2, V_2), it does not pass through intermediate equilibrium states and the process is clearly irreversible. It is important to realise that we cannot plot this irreversible process on an indicator diagram as the intermediate stages do not have well-defined pressures or temperatures; thus we represent an irreversible process schematically by a hatched line, as shown in Fig. 2.2. Additionally, the equation of state does *not* hold for the intermediate stages in an irreversible process.

We see that reversible processes are easy to recognise as there are never any finite pressure or temperature differences either within the system or between the system and the surroundings. Further, the direction of a reversible process can be changed by an infinitesimal change in the external conditions.

2.3 The bulk modulus and the expansivity

There are two physical properties of a system that are of great use in thermodynamics which are easy to measure: the *bulk modulus* K and the *volume thermal expansivity* β. A third quantity which is also easy to measure, the heat capacity, will be met in the next chapter.

The volume thermal expansivity, β, which is usually called simply the *expansivity*, is the fractional increase in volume divided by the temperature rise. In differential form,

$$\beta = \frac{1}{V}\left(\frac{\partial V}{\partial T}\right)_P \qquad [2.1]$$

where the subscript P outside the partial differential reminds us that the expansivity is measured at constant pressure and so strictly should be called the isobaric expansivity.

In one dimension, as for a wire, we have the *linear expansion coefficient*

$$\alpha = \frac{1}{L}\left(\frac{\partial L}{\partial T}\right)_{\mathscr{F}} \qquad [2.2]$$

α and β are related (see question 6, Chapter 2, Appendix 4):

$$\beta \approx 3\alpha$$

Moduli of elasticity are always given as stress/strain or force/unit area divided by the fractional deformation. For a solid or fluid, *the bulk modulus K* is given by

$$K = -V\left(\frac{\partial P}{\partial V}\right)_T = \frac{1}{\kappa} \qquad [2.3]$$

The constant T denotes that the bulk modulus is determined at constant temperature and so this is the isothermal bulk modulus. The negative sign ensures that K is a positive number because, for all known substances, dV is negative for a positive increase in pressure dP. The inverse of the bulk modulus is the *compressibility κ*.

For a stretched wire of cross-sectional area A, the appropriate modulus of elasticity is *Young's modulus*

$$Y = L/A\left(\frac{\partial \mathscr{F}}{\partial L}\right)_T \qquad [2.4]$$

(Note that, as dL is positive for positive $d\mathscr{F}$, we do not need a $-$ sign in front.)

2.4 An application of the thermodynamic method

Let us calculate the increase in tension $\mathscr{F}$ of a wire clamped between two rigid supports, a distance L apart, when it is cooled from T_1 to T_2 (Fig. 2.3).

We know that the equilibrium states of the wire are fixed by specifying two of the state variables $\mathscr{F}, L$ and T, which are related by some equation of state

$$g(\mathscr{F}, L, T) = 0$$

The wire undergoes a process in which it is changed from one equilibrium state $(\mathscr{F}_1, T_1)$ to another $(\mathscr{F}_2, T_2)$, both with the same length.

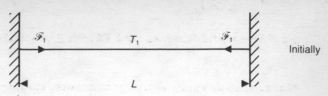

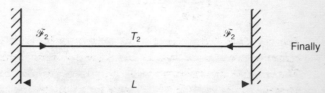

Fig. 2.3 A stretched wire being cooled under conditions of constant length.

Let us *suppose for a moment* that the wire is cooled from T_1 to T_2 *reversibly*. This could be achieved by bringing up to the wire a whole series of large bodies ranging in temperature from T_1 to T_2 to effect a quasistatic cooling through a sequence of equilibrium states.

For any one of these states, we may write

$$g(\mathscr{F}, L, T) = 0$$

or, solving for $\mathscr{F}$;

$$\mathscr{F} = \mathscr{F}(L, T)$$

where $\mathscr{F}(L, T)$ is a function of L and T alone. So

$$d\mathscr{F} = \left(\frac{\partial \mathscr{F}}{\partial T}\right)_L dT + \left(\frac{\partial \mathscr{F}}{\partial L}\right)_T dL$$

where the second term is zero as the cooling takes place under conditions of constant length.

Integrating,

$$\mathscr{F}_2 - \mathscr{F}_1 = \Delta \mathscr{F} = \int_{T_1}^{T_2} \left(\frac{\partial \mathscr{F}}{\partial T}\right)_L dT$$

Unfortunately, the integrand is $(\partial \mathcal{F}/\partial T)_L$ which we do not know. However, we do know

$$Y = L/A \left(\frac{\partial \mathcal{F}}{\partial L}\right)_T \quad \text{and} \quad \alpha = 1/L \left(\frac{\partial L}{\partial T}\right)_{\mathcal{F}}$$

which contain $\mathcal{F}$, L and T in different orders from the required $(\partial \mathcal{F}/\partial T)_L$. It is an easy matter to obtain this in terms of Y and α using the cyclical relation (see Appendix 2):

$$\left(\frac{\partial \mathcal{F}}{\partial T}\right)_L \left(\frac{\partial L}{\partial \mathcal{F}}\right)_T \left(\frac{\partial T}{\partial L}\right)_{\mathcal{F}} = -1$$

so $\quad \left(\dfrac{\partial \mathcal{F}}{\partial T}\right)_L = - \left(\dfrac{\partial \mathcal{F}}{\partial L}\right)_T \left(\dfrac{\partial L}{\partial T}\right)_{\mathcal{F}} = -YA\alpha$

Finally

$$\Delta \mathcal{F} = - \int_{T_1}^{T_2} YA\alpha \, dT = - YA\alpha \int_{T_1}^{T_2} dT$$

so $\quad \boxed{\Delta \mathcal{F} = - YA\alpha(T_2 - T_1)}$

if Y, A and α are independent of T. This is positive if $T_2 < T_1$, so the tension in the wire increases as expected.

But you may say 'This is fine if the cooling is *actually* reversible. In practice the cooling will not be reversible because we simply heat the wire and let it cool. This will result in large temperature gradients both within the wire itself and between the wire and the surroundings. As the intermediate states are not equilibrium states, we cannot apply the equation of state

$$\mathcal{F} = \mathcal{F}(L, T)$$

and our analysis appears to be invalid.'

The answer is that our analysis is still sound because the wire is being taken between *equilibrium states*. The initial tension of the wire is completely fixed by specifying the initial state (L, T_1), as is the final tension by specifying the final state (L, T_2). Thus the change in the tension is determined by specifying the end points:

$$\Delta \mathcal{F} = \mathcal{F}_2 - \mathcal{F}_1 = \mathcal{F}(L, T_2) - \mathcal{F}(L, T_1)$$

It does not matter how we go from state 1 to state 2 to determine the change in the state function $\mathscr{F}$. We say that $\Delta \mathscr{F}$ is *path independent*, being determined *only by the end points*. It is up to us to choose the most *convenient* path which happens to be a *reversible* path.

We use this elegant trick time and time again in thermodynamics to calculate changes in state functions for processes between a pair of equilibrium states.

2.5 Work

We conclude this chapter with a discussion of work, in preparation for the introduction of the first law of thermodynamics in the next chapter. To help us with our ideas, we shall initially consider our familiar gas-cylinder system (Fig. 1.1) and then generalise our conclusions to other systems.

Suppose that we have a gas in the initial equilibrium state (P_1, V_1) and we allow it to expand to a new equilibrium state (P_2, V_2) by decreasing the external balancing force on the piston and allowing this to slide out. If friction is present, some of the work the gas does in pushing the piston out is expended against these frictional forces; however, if no frictional forces are present, all of this work goes into performing work on the surroundings and it is possible to find a very simple expression for the work done in terms of the state variables of the gas, provided that the expansion is performed quasistatically so that the pressure is well defined and uniform throughout the gas. In other words, we have to expand the gas reversibly.

Suppose that, at one of the intermediate equilibrium states during the reversible expansion, the pressure is P and the balancing force on the piston is $\mathscr{F}$. Then, as shown in Fig. 2.4

$$\mathscr{F} = PA$$

where A is the area of the piston. If the force is decreased infinitesimally so that the piston moves out by dx, the work done by the gas against the surroundings applying the force $\mathscr{F}$ is

$$dW = PA\,dx = P\,dV \qquad \text{(reversible)} \qquad [2.5]$$

The total work performed in the process is

$$W = \int_{V_1}^{V_2} P\,dV \qquad\qquad \text{(reversible)} \qquad [2.6]$$

where we have written reversible after these equations to remind us that they are true for reversible processes.

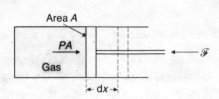

Fig. 2.4 The work done by a gas when it expands reversibly through an infinitesimal volume change, dV, is $P.dV$.

This reversible work is in fact the maximum work that can be obtained from the expansion. We can see this as follows. The work done by the expanding gas against the surroundings is $\int_{V_1}^{V_2} \mathscr{F}\,dx$ and this will be a maximum when $\mathscr{F}$ is as large as possible at all stages in the expansion. The largest possible value for $\mathscr{F}$, if the gas is to expand, is infinitesimally less than PA and this leads to equation [2.6] again for the maximum work.

Although equations [2.5] and [2.6] apply to reversible processes, they also apply to *some* irreversible processes too where the actual expansion (or compression), when considered by itself, is quasistatic but where there is irreversibility elsewhere in the system. This is best illustrated by two examples.

1. Let us consider a cylinder equipped with a frictionless piston containing two solids which react slowly to produce a gas, as in Fig. 2.5(a). Because the gas is released slowly, the pressure P inside the cylinder is only infinitesimally greater than the pressure P_0 of the surroundings and the piston is always infinitesimally close to mechanical equilibrium. The chemical reaction is irreversible in that it cannot be reversed by an *infinitesimal* change in the external conditions, such as the

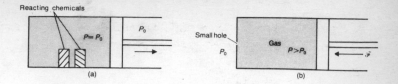

Reacting chemicals

$P = P_0$

P_0

(a)

Small hole

P_0

Gas $P > P_0$

$\mathscr{F}$

(b)

Fig. 2.5 Two examples of $\mathrm{d}W = P.\mathrm{d}V$ when the process is irreversible.

pressure or the temperature. By the argument that we have just used, the work done by the system against the surroundings is again

$$W = \int_{V_1}^{V_2} P\mathrm{d}V = \int_{V_1}^{V_2} P_0\mathrm{d}V = P_0(V_2 - V_1) \quad [2.7]$$

where V_1 and V_2 are the initial and final volumes, even though the whole process is irreversible because of the chemical reaction.

2. Consider as a second example the pump, shown in Fig. 2.5(b), containing gas at the high pressure P. The frictionless piston is pushed in slowly, thus expelling the enclosed gas through the small hole at the end into the surrounding atmosphere at the lower presure P_0. As the piston is being pushed in slowly, the piston is always infinitesimally close to being in mechanical equilibrium, with the applied force being only infinitesimally greater than PA. However, the whole process is irreversible because, even if $\mathscr{F}$ is reduced slightly, the process will not stop, and gas will still flow out through the small hole. The work done by the surroundings on the gas is again

$$W = - \int_{V_1}^{V_2} P\mathrm{d}V$$

where V_1 and V_2 are the initial and final volumes. The negative sign has been inserted to make this work a positive number; this point will be discussed in section 2.7 below.

Let us now return to our discussion of reversible processes. The most convenient way to represent the expansion of a gas is on an indicator diagram, where it is seen from equation [2.6] that the work done in a reversible process is the area under the curve or path for the process. As there are an infinite number of paths connecting 1 and 2, the work done depends on the actual path chosen, i.e. on the way P varies with V.

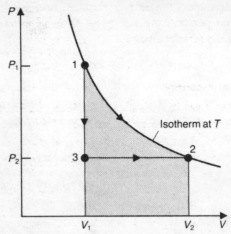

Fig. 2.6 Work depends on the path.

For example, suppose our expansion were isothermal, as would be the case if the cylinder walls were diathermal and were in contact with a thermal reservoir at T. This path is represented by the upper curve 1–2 shown in Fig. 2.6. Then

$$W = \int_{V_1}^{V_2} P\,dV = nRT \int_{V_1}^{V_2} dV/V = nRT\ln(V_2/V_1)$$

[2.8]

as $PV = nRT$.

Another simple reversible path is 1–3–2 in Fig. 2.6, consisting of an isochoric (constant volume) decrease of pressure 1–3 followed by an isobaric expansion 3–2. For this process, the work is simply $P_2(V_2 - V_1)$, which is different from that for the isothermal expansion.

We conclude then that work in general is path dependent and it cannot be expressed simply as the difference between the two end point values of some state function. This is in contrast with the volume, say, which is uniquely defined by the state of the system and where the change is always $V_2 - V_1$, irrespective of the process used to take the system from 1 to 2. In the language of Appendix 2, we write the infinitesimal work term as $đW$ where the bar through the d denotes that $đW$ is an *inexact differential*. However, it should be remarked that, while in general, work is path dependent, there is a class of processes where the work is path independent; we shall meet these in the next chapter when we discuss adiabatic work.

2.6 The free expansion

Consider a gas in the state (P, V) being contained in the left-hand part of a double-sectioned chamber as in Fig. 2.7. There is a vacuum in the right-hand part. For simplicity, we shall take the volumes of each part as being equal, at V. Let the intervening partition be broken so that the gas rushes into the right-hand half before eventually settling down to a new equilibrium state. This process is known as a free expansion. (Strictly, the walls should also be adiabatic for a true free expansion, but this need not concern us here.)

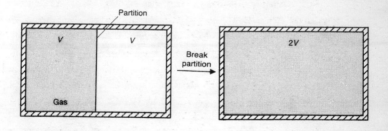

Fig. 2.7 A free expansion. When the partition is broken, the gas occupies the whole volume but no work is performed on the surroundings.

How much work is done by the gas in this process? A blind application of $W = \int_{V_1}^{V_2} P dV$ would give a finite answer, as the volume certainly changes. Of course, the answer is zero as the gas does no work on the surroundings outside the chamber. The expression $W = \int P dV$ cannot be applied here as this process is not reversible. This example illustrates two points in thermo-dynamics:

1. It is important to be clear as to what is the system. Here it is the chamber as a whole and not just the left-hand part initially containing all the gas.
2. $đW = PdV$ is applicable only to reversible processes and to those irreversible processes where the expansion is frictionless and quasistatic, such as those considered in section 2.5.

2.7 The sign convention for work

There are unfortunately different sign conventions used for W in different texts. We shall adopt the convention that, when the sur-roundings do work on the system, that work is positive; con-versely, when the system does work on the surroundings, that work is negative. We have seen that, if the gas system expands against the surroundings, the infinitesimal work done is $đW = PdV$. As it stands, this is positive for an expansion, as dV is then posi-tive. To make $đW$ a negative quantity for an expansion, we re-write equation [2.5] as

$$\boxed{đW = -PdV}$$
(reversible)   [2.9]

If on the other hand we, as the surroundings, compress the gas, dV is then negative and $đW$ is positive as required.

In choosing this convention, we are bearing in mind that, as physicists, we regard the system as being the important thing. The other convention, which is usually adopted by engineers, regards positive work as that done *by the system* on the surroundings, with $đW = +PdV$. Engineers are interested, equally rightly, in how much work they can get out of the system, and it is this work which is important for them.

Whichever convention is chosen, it makes no difference to the fundamental thermodynamic relations which we shall develop, but one must stick consistently with one or the other.

2.8 Dissipative work

Suppose we have a viscous fluid which can be stirred, as in Fig. 2.8, by the action of the falling weight. Because of the dissipative viscous effects in the fluid, the temperature will rise and the state of the fluid system will change. This work is performed irreversibly: if the torque on the shaft attached to the stirrer were reduced infinitesimally, the shaft would not start to go the other way with the weight rising up again. We call this irreversible work *stirring work* or simply *dissipative work*. Another example of dissipative work is the passing of a current I through a resistor R immersed in the fluid. Then the work performed in the time t is I^2Rt. Again this work is irreversible because reducing the battery voltage slightly will not cause the current to reverse, with the resistor then doing work on the battery.

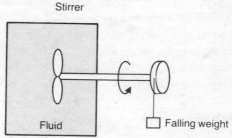

Fig. 2.8 Dissipative work.

Unlike reversible work considered in section 2.5, it is not possible to find an expression for dissipative work in terms of the state variables of the system.

2.9 Other kinds of work

There are systems other than our compressible gas or fluid which are commonly encountered in thermodynamics, and we need to examine the appropriate forms for the infinitesimal work term in a reversible process for each of the following.

An extensible wire

The work done by us acting as the surroundings when we stretch a wire at a tension $\mathcal{F}$ through an infinitesimal distance dx is

$$ dW = \mathcal{F}dx \qquad\qquad [2.10] $$

This is positive for a positive extension dx, which is consistent with our sign convention.

A surface film

A surface film, such as that in a soap bubble, has equilibrium states which are completely specified by the state variables, the area A and the surface tension Γ. Normally Γ is found experimentally to depend on the temperature only and not on the area.

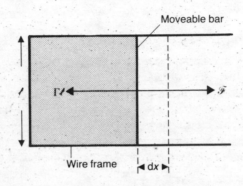

Fig. 2.9 The work done in increasing the surface area of a film by dA is Γ dA.

Consider the film shown in Fig. 2.9 being stretched isothermally, with the moveable bar being pulled an infinitesimal distance dx by an external force which is only infinitesimally greater than $\Gamma \ell$. Then, the infinitesimal work done is

$$đW = \Gamma \ell dx = \Gamma dA \qquad [2.11]$$

This is positive if dA is positive. (We have assumed a single-sided film here.)

A reversible electrolytic cell

The equilibrium states of a reversible electrolytic cell, such as the simple Daniell cell, are specified by the state variables, the charge Z stored and the emf $\mathcal{E}$. Suppose that Z is increased infinitesimally by dZ; then the work performed by the external charging circuit is

$$\text{d}W = \mathcal{E}\,\text{d}Z \qquad\qquad [2.12]$$

A comment should be added here about our notation. We have had to use the symbol Z, instead of the more usual symbol Q, for charge because, in thermodynamics, the symbol Q is reserved for heat. As we shall see in section 8.11 both Z and Q appear in the thermodynamics of an electrolytic cell, and it is important to make a distinction.

A simple magnetisable material

The equilibrium states of a simple magnetic material are specified by the state variables, the overall magnetic moment $\mathcal{M}$ and the applied magnetic induction B_0. Paramagnetic and diamagnetic compounds fall into this category, but we have to exclude most ferromagnetics where hysteresis effects result in there being no unique relation between $\mathcal{M}$ and B_0 at each temperature. If the magnetisation is uniform over the volume of the sample,

$$\mathcal{M} = MV$$

where M is the magnetisation, or the magnetic moment per unit volume.

In Appendix 3 we show that, when the sample is uniformly magnetised, the external work required to increase the magnetic moment from $\mathcal{M}$ to $\mathcal{M} + \text{d}\mathcal{M}$ in the applied induction field B_0 is

$$\text{d}W = B_0\,\text{d}\mathcal{M} \qquad\qquad [2.13]$$

A dielectric material

The equilibrium states of a dielectric substance are specified by the state variables, the overall electric dipole moment $\mathcal{P}$ and the applied electric field E. It is also shown in Appendix 3, where we consider only linear dielectrics with no hysteresis effects, that the infinitesimal external work required to increase the overall dipole moment of a uniformly polarised dielectric from $\mathcal{P}$ to $\mathcal{P} + \text{d}\mathcal{P}$ in a field E is

$$dW = Ed\mathscr{P} \qquad [2.14]$$

For such a uniformly polarised dielectric, the overall dipole moment is related to the polarisation P, or dipole moment per unit volume, by

$$\mathscr{P} = PV$$

We have collected all these results together in Table 2.1:

Table 2.1   *Infinitesimal work in various reversible processes*

System	Intensive variable	Extensive variable	Infinitesimal work
Gas or fluid	P	V	$-PdV$
Film	Γ	A	ΓdA
Cell	$\mathscr{E}$	Z	$\mathscr{E} dZ$
Magnetic material	B_0	$\mathscr{M}$	$B_0 d\mathscr{M}$
Dielectric material	E	$\mathscr{P}$	$Ed\mathscr{P}$

Suppose we were to consider as our system only *part* of our original system. Then, if our system were the usual gas, the pressure of the subsystem considered would be the same as in the original system but the volume would be smaller. Because the pressure is in this sense size independent, we say that it is an *intensive variable*; conversely, volume is an *extensive variable*. In Table 2.1 the state variables are grouped according to whether they are extensive or intensive.

There is one final point. *All* the work processes that we have considered may be said to be equivalent to a process whose sole effect on the surroundings is the raising or lowering of a weight, and so may be said to be equivalent to mechanical work. Notice that a weight does not actually have to be raised, only that it *could* be raised. Clearly this is so for the case of an expanding gas, as shown in Fig. 2.10(a). As another example, we may consider dissipative electrical work where a current I enters the system containing a resistance R, as in Fig. 2.10(b). The current may be thought of as being produced by a generator, the shaft of which is turned by the action of a falling weight. Work is then done on the system at the rate of I^2R.

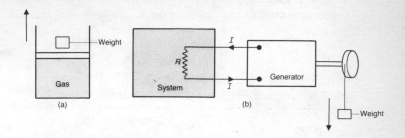

Fig. 2.10 Work is always equivalent to the raising of a weight.

Similar processes involving the lifting or lowering of a weight may be thought up for all the other forms of work considered. We shall find this idea useful when we come to differentiate between work and heat in the next chapter.

2.10 An example of the calculation of work in a reversible process

Let us illustrate these ideas by calculating the work done in changing the state of a compressible fluid from (P_1, T_1) to (P_2, T_2) in a reversible process.

The infinitesimal work done in part of the process is $-P\mathrm{d}V$, so we need to find $\mathrm{d}V$. This example is typical of many in thermodynamics in that one has to find the change in one state function when one is told the change in two others – here $(P_2 - P_1)$ and $(T_2 - T_1)$. The technique is always the same:

Write the equation of state in the form that gives the state function whose change we wish to find in terms of the other two whose changes are given.

Hence writing

$$V = V(P, T)$$

$$dV = \left(\frac{\partial V}{\partial P}\right)_T dP + \left(\frac{\partial V}{\partial T}\right)_P dT$$

As $K = -V\left(\frac{\partial P}{\partial V}\right)_T$ and $\beta = \frac{1}{V}\left(\frac{\partial V}{\partial T}\right)_P$

$$dV = -\frac{V}{K} dP + \beta V dT$$

so $$ðW = \frac{PV}{K} dP - P\beta V dT$$

and $$W = \int_{P_1}^{P_2} \frac{PV}{K} dP - \int_{T_1}^{T_2} P\beta V dT$$

In essence, we have solved the problem if we can perform the integrations and this we can do in certain simplified cases. Let us keep the change isothermal for example. Then

$$W = \int_{P_1}^{P_2} \frac{PV}{K} dP = V(P_2^2 - P_1^2)/2K$$

if the volume stays approximately constant during the process.

Chapter 3
The first law of thermodynamics

The first law of thermodynamics is essentially a 'balance sheet of energy'. It gives the precise relationship between the familiar concept of work and the new concepts of internal energy and heat, both of which we shall be defining shortly.

It is interesting to approach the first law historically.

3.1 Joule's experiments

At the beginning of the nineteenth century, the dominant theory as to the nature of heat was that it was an indestructible substance (*caloric*) which flowed from a hot body, rich in caloric, to a cold body which had less caloric. Heat was quantified by the temperature rise it produced in a unit mass of water, taken as a standard reference substance. The experiments of Black at the end of the eighteenth century had shown that, when two bodies were put in thermal contact, the heat lost by one in this 'method of mixtures' was equal to the heat gained by the other. This seemed to confirm that heat was a conserved entity.

However, when Benjamin Thompson (who became Count Rumford) was working in the arsenal in Munich supervising the boring of cannon, he noticed that great heat was produced, as measured by the temperature rise in the cooling water. Further, when he used a blunt boring tool, he found that he could even *boil* the water, with the supply of heat being apparently inexhaustible. He concluded that heat could not be a finite substance such as caloric and that there was a direct relation between the work done and the heat produced. The precise relation was established by Joule some fifty years later in a careful series of experiments between 1840 and 1849.

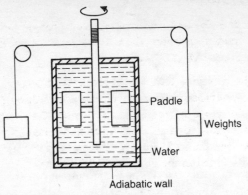

Fig. 3.1 A schematic representation of Joule's apparatus.

Joule, a Manchester brewer, constructed a tub containing a paddle wheel which could be rotated by the action of weights falling outside the tub, as in Fig. 3.1. Water inside the tub could thus be stirred (irreversibly because of turbulence), raising its temperature between two equilibrium states. The walls of the tub were insulating, so the work was performed under adiabatic conditions; we call such work *adiabatic work*. Working with extraordinary accuracy, Joule found the following:

1. That it required 4.2 kJ of work to raise the temperature of one kg of water through one degree Celsius (we have converted his British units to the modern SI units). This is known as the *mechanical equivalent of heat, J*. It is interesting to note that, when Joule examined Rumford's results, he obtained a value for J that was consistent with his own.

2. That no matter how the adiabatic work was performed, it always required the *same* amount of work to take the water system between the same two equilibrium states. Joule varied his adiabatic work by changing the weights and the number of drops. He also performed the same amount of adiabatic work electrically by allowing the current produced by an electrical generator to be dissipated in a known resistance immersed in the water.

3.2 The first law of thermodynamics and the internal energy function

We may summarise these findings in the following statement of the first law:

If a thermally isolated system is brought from one equilibrium state to another, the work necessary to achieve this change is independent of the process used.

This is saying that the adiabatic work $W_{adiabatic}$ expended in a process is path independent, depending only on the end equilibrium points; and this is true whether or not the process is reversible. So there must exist a state function whose difference between the two end points 2 and 1 is equal to the adiabatic work. We call this state function the *internal energy U* with

$$W_{adiabatic} = U_2 - U_1 \qquad [3.1]$$

In mechanics, we are familiar with the idea of the work done on a system increasing the kinetic and potential energies. In our discussion we have *excluded* any change in these bulk energies: Joule's tub was neither lifted up nor was it set in motion across the floor of the laboratory. From a molecular viewpoint, however, the external work does in fact go into increasing kinetic and potential energies — those of the individual molecules which have kinetic energy because of their *random* motion and potential energy because of their mutual attraction. This viewpoint is helpful in understanding the physical meaning of internal energy.

3.3 Heat

If the system is not thermally isolated, it is found that the work W done in taking the system between a pair of equilibrium points depends on the path. Now, for a given change, $\Delta U = U_2 - U_1$, is fixed but W is not now equal to ΔU. In other words there is a difference between the adiabatic work required to bring about a change between two equilibrium states and the non-adiabatic work required to effect the same change, with the latter having an infinite number of possible values.

We call the difference between ΔU and W the *heat*, Q. The generalisation of equation [3.1] is then

$$\boxed{\Delta U = W + Q}$$ [3.2]

which is the mathematical statement of the first law. It tells us that the internal energy can be increased either by doing work on or by supplying heat to the system. In this form, it is true for all processes, whether reversible or irreversible. Now we have seen that all forms of work are equivalent to the mechanical raising or lowering of a weight in the surroundings so we conclude that

> *Heat is the non-mechanical exchange of energy between the system and the surroundings because of their temperature difference.*

(Here we are considering closed systems. The transfer of energy in open systems is discussed in Chapter 10.)

There has to be a sign convention for heat, just as we have one for work. We define Q as positive when it enters the system, so equation [3.2] is correct as it stands, with U increasing if we do work on the system and if we allow heat to flow in.

For an infinitesimal process, the first law takes the form

$$dU = đW + đQ$$

where we write both $đQ$ and $đW$ with bars through them to indicate that W, and therefore Q, are in general path dependent. In the language of Appendix 2, we say they are *inexact differentials*. Although in the special case of adiabatic work $\int đW_{adiabatic}$ is path independent and in that sense $đW_{adiabatic}$ is an exact differential, we shall consistently write the infinitesimal work term as $đW$ with a bar for all cases, as W is not a state function.

If we have a compressible fluid, where $đW = -PdV$ for an infinitesimal reversible process, the first law becomes

$$dU = -PdV + đQ$$

or $$\boxed{đQ = dU + PdV}$$ (reversible) [3.3]

It is very important to realise that this form applies to a reversible infinitesimal process, and so we have added reversible in brackets to remind us.

Let us give an example which illustrates the difference between heat and work. In Fig. 3.2 we have a gas in a container with rigid diathermal walls. A current I flows through the heating coils of resistance R wrapped round the container. In Fig. 3.2(a) the system, denoted by the dotted line, includes the heating coils. Now we know, from our discussion at the end of section 2.9, that work is being done on the system at the rate I^2R because the current I enters the system. The energy crossing the system boundary is in the form of *work*. In Fig. 3.2(b) the system is the gas and container alone, excluding the coils. Here, no work is done on the system but there is energy flow across the system boundary in the form of *heat* because the temperature of the coils is higher than that of the gas. This simple example shows that, in differentiating between heat and work, it is very important to be clear as to what constitutes the system.

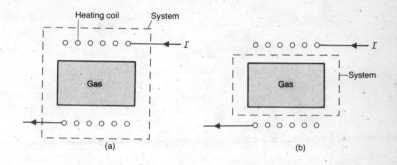

Fig. 3.2 An illustration of the difference between work and heat. The input of energy into the system is in (a) as work and in (b) as heat.

Finally, it is helpful to differentiate between heat and work from a microscopic viewpoint. When we add energy to the system in the form of heat, we increase the *random* motion of the constituent molecules. However, when we increase the energy by performing work, we displace the molecules in an *ordered* way. Consideration of the stretching of a spring immediately illustrates this point. Work then results in organised motion while heat results in random motion. It is interesting to apply these ideas to a gas in a cylinder and to see how a rise in temperature can be achieved either by the addition of heat or the performance of work. If heat is added through diathermal walls, this increases the random kinetic energy of the molecules, which means a rise in temperature. Why is there a rise in temperature if work is performed on the gas which we now consider to be in a cylinder with adiabatic walls? When the piston is pushed in, the molecules striking the piston are accelerated in the direction of its travel. If these molecules strike the walls of the cylinder, they cannot lose energy to the surroundings because the walls are adiabatic. However, any organised motion initially imparted to these molecules is rapidly randomised by collisions, either with the walls or with other molecules. This increase in the random kinetic energy appears again as a temperature rise.

3.4 Heat capacity

Suppose we have a process where we allow heat Q to flow into a system, changing it from one equilibrium state to another with a temperature difference ΔT, as in Fig. 3.3. The *heat capacity* C of a system is defined as the limiting ratio of the heat introduced reversibly into the system divided by the temperature rise:

$$C = \underset{\Delta T \to 0}{\text{Limit}} (Q/\Delta T) = \text{đ}Q/\text{d}T \qquad [3.4]$$

The *specific heat* c is the heat capacity per unit mass:

$$c = 1/m \ \text{đ}Q/\text{d}T \qquad [3.5]$$

(We shall use lower-case symbols frequently in this book when we wish to denote specific quantities which are quantities referred to unit mass, which could be a kilogram or a mole.)

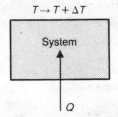

$T \to T + \Delta T$

System

Q

Fig. 3.3 The heat capacity is $\underset{\Delta T \to 0}{\text{Limit}} Q/\Delta T$.

Now a process is not completely defined simply by the temperature difference ΔT between the end points. There are a large number of possible reversible paths between these end points, each with a different Q. It follows that there are a large number of possible heat capacities. We shall restrict ourselves to two.

1. *The heat capacity at constant volume, C_V*

Suppose we heat the system under conditions of constant volume. For an infinitesimal isochoric reversible process, equation [3.3] gives $đQ_V = dU$ where the subscript V on $đQ_V$ indicates that the volume is held constant. Although $đQ$ for this special case is equal to the exact differential dU, we shall persist with the notation of keeping a bar written through because Q is not a state function. Thus

$$C_V = \frac{đQ_V}{dT} = \left(\frac{\partial U}{\partial T}\right)_V \qquad [3.6]$$

2. *The heat capacity of constant pressure, C_P*

Let us now heat the system at constant pressure. The heat capacity at constant pressure is then

$$C_P = \frac{đQ_P}{dT}$$

where $đQ_P$ is the heat added reversibly and isobarically to produce the temperature rise dT. Just as we found that $đQ_V$ was equal to dU, we can obtain a similar result for $đQ_P$ by defining a new state function, the *enthalpy H*, as

$$H = U + PV \qquad [3.7]$$

so $\quad dH = dU + PdV + VdP$

for an infinitesimal process. Using the infinitesimal form of the first law, equation [3.3], we have

$$dH = đQ + VdP \qquad [3.8]$$

which holds again for a reversible process. For a reversible, isobaric process then,

$$dH = đQ_P \qquad [3.9]$$

That is, the heat evolved in a reversible isobaric process is equal to the enthalpy change. Thus

$$C_P = \frac{\text{d}Q_P}{\text{d}T} = \left(\frac{\partial H}{\partial T}\right)_P \qquad [3.10]$$

We shall shortly be meeting an engineering use for H when we consider steady flow processes and deal with turbines. Enthalpy also has a particular use in chemistry; it will be shown in Chapter 6 that the enthalpy change in an isobaric chemical reaction is equal to the *heat of reaction*, whether or not that reaction is reversible.

Let us return to our definition of heat capacity and the requirement that the heat has to be introduced reversibly. This means that, as the heat is introduced, the system must pass through a series of equilibrium states, with the pressure and temperature always being uniform throughout the system. What would be the consequences of putting in a small burst of heat at some point in the system, which we shall, for example, take to be a gas, from a source at a finite temperature above that of the system? This will cause local heating, with pressure and temperature gradients being set up in the gas system and we hardly have the conditions required for a determination of the heat capacity. In practice, heat capacities are measured in just this way and there does appear to be an inconsistency. If, however, the relaxation time for the attainment of an equilibrium state in the system is much shorter than the time scale of the heating, the system is always so close to an equilibrium state that there are no significant internal pressure and temperature gradients; the irreversible nature of the heating is then of *no* consequence. In fact, this is the usual situation and so there is no inconsistency between our definition of heat capacity and its experimental determination.

3.5 The method of mixtures

At the beginning of this chapter we mentioned that Black, who was actually a professor of medicine, had quantified heat using the method of mixtures. Let us see how the first law justifies the idea of 'heat lost equals heat gained'.

Suppose we have two systems, A and B, put in thermal contact with an adiabatic wall surrounding both systems. For the two systems, the first law is

$$U_f^A - U_i^A = Q^A + W^A$$

$$U_f^B - U_i^B = Q^B + W^B$$

where i and f refer to the initial and final states and the superscripts A, B refer to the different systems. Adding,

$$(U_f^A + U_f^B) - (U_i^A + U_i^B) = (Q^A + Q^B) + (W^A + W^B)$$

Now $(U_f^A + U_f^B) - (U_i^A + U_i^B)$ is the change in the internal energy of the *composite* system and $(W^A + W^B)$ is the work done on it. Hence $(Q^A + Q^B)$ is the heat that flows into the composite system, which we know to be zero as this composite system is surrounded by an adiabatic wall. So

$$Q^A + Q^B = 0 \text{ and } Q^A = -Q^B$$

or 'heat lost equals heat gained'.

3.6 Ideal gases

Some simple results can be obtained from the first law of thermodynamics for the particular case of ideal gases. We have previously defined an ideal gas as one which obeys the equation of state

$$PV = nRT$$

We shall show in section 7.3, as a consequence of this equation of state, that the internal energy is a function of temperature alone. This is consistent with the result from kinetic theory that the kinetic energy per mole is $3/2\, N_A k_B T$ where N_A is Avogadro's number and k_B is the Boltzmann constant.

If we write then

$$U = U(T) \qquad [3.11]$$

we obtain the following three important results for an ideal gas.

The free expansion

We met the concept of a free expansion in the previous chapter where we saw that no external work against the surroundings was performed when the gas expanded irreversibly into the whole of the chamber. Suppose the walls are adiabatic so no heat enters the system. As $W = 0$ and $Q = 0$, the first law shows that $U_i = U_f$, where we use the subscripts i and f to denote the initial and final equilibrium states. As $U = U(T)$, $T_i = T_f$ and so there is *no temperature change between the end states*.

In 1843, Joule attempted to measure the temperature change in a free expansion for air. He was unable to detect, within the experimental error, any temperature change. With modern sensitive equipment, we are able to measure this and it is found to be very small; thus air approximates at normal temperatures to an ideal gas. It is found that *all* known gases cool slightly on undergoing a free expansion. This is consistent with the kinetic theory idea that temperature is associated with the kinetic energy of the molecules. If the gas expands, then the intermolecular attraction potential energy goes up as the molecules get further apart. As the total internal energy U is constant for the free expansion, this means that the kinetic energy, and therefore the temperature, goes down.

The quantity $(\partial T/\partial V)_U$ is a measure of the cooling effect occurring in a free expansion and is known as the Joule coefficient μ_J. Later, we shall see how thermodynamics helps us to derive an expression for μ_J from the equation of state, even though a free expansion is an irreversible process.

$C_P - C_V$

Let us consider n moles of an ideal gas. We have $đQ = dU + PdV$ for an infinitesimal reversible process. As $C_V = (\partial U/\partial T)_V$ and, for the special case of an ideal gas where $U = U(T)$, we have the simple relation

$$C_V = \frac{dU}{dT} \quad \text{or} \quad dU = C_V dT \qquad \text{(ideal gas)}$$

The infinitesimal form of the first law becomes

$$đQ = C_V dT + P dV$$

We now consider a constant pressure process. Then, dividing all through by dT and taking the partial differential at constant P

$$\frac{đQ_P}{dT} = C_P = C_V + P\left(\frac{\partial V}{\partial T}\right)_P$$

As $PV = nRT$ so that $P(\partial V/\partial T)_P = nR$, it follows that

$$\boxed{C_P = C_V + nR} \tag{3.12}$$

This relation between C_P and C_V is very simple for an ideal gas; we shall derive a relation for $C_P - C_V$ for a general system in section 7.1.

The equation of an adiabatic

We know that, in an isothermal reversible expansion at temperature T, the pressure and volume of an ideal gas are always related by $PV = nRT$. There is a very simple relation between P and V if the expansion is performed both *adiabatically* and *reversibly*. We have

$$đQ = dU + P dV \qquad \text{(reversible)}$$

so $\quad 0 = dU + P dV \qquad$ (reversible and adiabatic)

or $\quad 0 = C_V dT + P dV$ as $dU = C_V dT$

But the equation of state $PV = nRT$ holds at all points in this expansion also, although T is not now constant as it was for the isothermal expansion. We may substitute for P from the equation of state in our previous expression to obtain

$$0 = C_V dT + nRT\frac{dV}{V}$$

or $\quad 0 = C_V\frac{dT}{T} + nR\frac{dV}{V}$

Integrating

$$C_V \ln T + nR \ln V = \text{a constant*}$$

or, dividing all through by C_V and using equation [3.12],

$$\ln T + (\gamma - 1) \ln V = \text{a constant*}$$

where the ratio of the heat capacities C_P/C_V is written as γ. Hence

$$TV^{\gamma - 1} = \text{a constant*}$$

Using the equation of state again, we have finally that

$$PV^{\gamma} = \text{a constant*} \tag{3.13}$$

which is our equation for a *reversible adiabatic*.

It is a simple matter to show by differentiation that the slope of the adiabatic at a particular point (P, V) on the $P - V$ indicator diagram is γ times that of the isotherm through that point. The result of the previous section shows that $\gamma > 1$ as nR is a positive number so the adiabatic has the steeper slope. This is shown in Fig. 3.4.

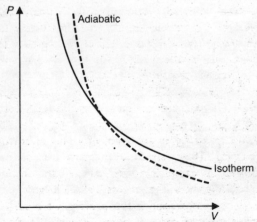

Fig. 3.4 The adiabatic for an ideal gas has a slope γ times that for the isotherm at each point on the $P - V$ indicator diagram.

* The constants here are all different.

3.7 Non-ideal gases

In a real gas, there are intermolecular attractions. Additionally, the molecules themselves occupy a finite volume. The equation of state therefore has to be modified from the simple $PV = nRT$ for an ideal gas. The most successful modification is that of van der Waals. His equation is

$$(P + a/v^2)\,(V - nb) = nRT \qquad [3.14]$$

where the first bracket on the left contains the modification a/v^2 to the pressure because of the molecular interactions and the second bracket contains the modification nb to the volume to take into account the molecular volume (a and b are constants). The reader is referred to other texts for a discussion of the a/v^2 term. Notice that the first bracket contains the molar volume $v = V/n$ and not the total volume V; otherwise, the left-hand side would not be extensive.

3.8 The Joule–Kelvin effect

Although we shall discuss general flow processes in the next section, and shall return again to this particular flow process in section 7.7, we shall find it very instructive to formulate some of our ideas at this stage. The effect is used in the liquefaction of gases and is often called, for reasons that will be immediately obvious, the *throttling process*.

The process is illustrated in Fig. 3.5(a). Gas is forced at a constant pressure and at a steady rate through a small hole, or series of holes, to emerge at a constant pressure. The series of small holes is usually in the form of a plug of cotton wool or similar material. There is a finite pressure drop across the plug, so the process is irreversible. The walls of the chamber are thermally insulating and so the process is also adiabatic.

In order to analyse this process, we focus our attention on a given mass of gas as it passes through the plug. We consider the entirely equivalent situation depicted in Fig. 3.5(b) of this gas being contained in the left-hand cylinder and slowly being forced at constant pressure P_i through the plug. As the gas enters the right-hand cylinder, the piston there moves out to maintain the

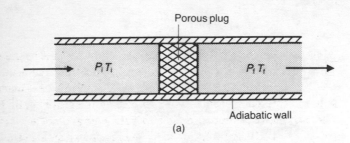

Porous plug

$P_i T_i$

$P_f T_f$

Adiabatic wall

(a)

$P_i T_i$
V_i

P_f

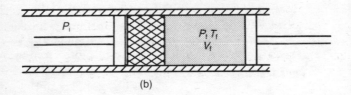

P_i

$P_f T_f$
V_f

(b)

Fig. 3.5 A schematic representation of the throttling process.

48   *The first law of thermodynamics*

pressure at the constant value P_f. Let us consider all of the gas being in the initial equilibrium state (P_i, V_i) and then being pushed through the plug to the right-hand side where it reaches the equilibrium state (P_f, V_f). Because there is no pressure drop across the left-hand piston and as there is no friction, the work we do in forcing the gas system through the plug is

$$ W = - \int_{V_i}^0 P_i \mathrm{d}V = P_i V_i $$

Although the process is irreversible, we have been able to use the $-P\,\mathrm{d}V$ expression for the work done because the argument we used in our original derivation in Chapter 2 is valid here — namely that the external force necessary to push in the piston is still PA. The only finite pressure drop is *across the plug* and not *across the piston*. Similarly, the work done by the gas on expanding into the right-hand cylinder is $P_f V_f$.

Applying the first law to the gas,

$$ U_f - U_i = 0 + P_i V_i - P_f V_f $$

as no heat enters. So

$$ U_i + V_i P_i = U_f + V_f P_f $$

or

$$ \boxed{H_i = H_f} $$ [3.15]

The throttling process is thus as *isenthalpic* one.

The quantity $(\partial T/\partial P)_H$ is a measure of the temperature change occuring in a throttling process and is known as the Joule–Kelvin coefficient μ_{JK}. Later, we shall use thermodynamics to calculate μ_{JK} even though the process is irreversible. Unlike the Joule expansion, where there is *always* cooling, *both* heating and cooling can occur in the throttling process.

3.9 Steady flow processes — the turbine

We conclude this chapter with a brief discussion of steady flow processes, which are of particular importance in engineering. By such a process we mean the flow of a fluid at a constant rate through a device so that some of the internal energy of the fluid is transformed into mechanical work. The device, for example, might be an air compressor, a refrigerator or a turbine. In Fig. 3.6 we depict a general steady flow process.

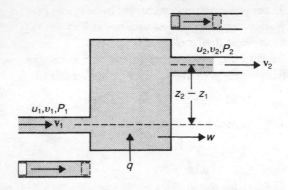

Fig. 3.6 A schematic representation of a general steady flow process.

Let us focus our attention on a unit mass of the fluid flowing through the device. We treat this unit mass as the system. The relevant parameters, which are specific values for the extensive quantities in that they refer to unit mass, are listed in Table 3.1.

Table 3.1   *Relevant parameters in a steady flow device*

	Entering device	*Leaving device*
Pressure	P_1	P_2
Volume	v_1	v_2
Height	z_1	z_2
Flow velocity	v_1	v_2
Internal energy	u_1	u_2

In order to help us to determine the work done by the fluid, let us imagine our unit mass of fluid being contained in a cylinder and being forced at the constant pressure P_1 into the device, just as in the throttling process. The work done on the fluid is then P_1v_1, as before. Similarly the work done by the fluid on emerging from the device is P_2v_2. Let the device perform, in addition, the work w (by w we mean, for example, the shaft work done by a turbine). Let heat q enter the system. We can now summarise all the energy changes, the work performed and the heat flow:

1. the internal energy changes by $u_2 - u_1$;
2. the kinetic energy changes by $\frac{1}{2}(v_2^2 - v_1^2)$;
3. the potential energy changes by $g(z_2 - z_1)$;
4. the net work done *on* the fluid is $P_1 v_1 - P_2 v_2 - w$;
5. the heat flow *into* the system is q.

Because the bulk potential and kinetic energies are changing in this process, we have to modify the first law to

$$\Delta(KE + PE) + \Delta U = W + Q$$

Substituting all these values,

$$\frac{1}{2}v_2^2 - \frac{1}{2}v_1^2 + g(z_2 - z_1) + u_2 - u_1$$
$$= P_1 v_1 - P_2 v_2 - w + q$$

Remembering that the specific enthalpy $h = u + Pv$, we have for the shaft work

$$\boxed{w = h_1 - h_2 + \frac{1}{2}(v_1^2 - v_2^2) + g(z_1 - z_2) + q} \qquad [3.16]$$

This is the general energy equation for steady flow.

The values for h at different temperatures and pressure are tabulated for different substances in 'engineering heat tables' and we can immediately compute w for different flow systems. We shall consider just two important constant flow processes.

The turbine

Although the temperature of a gas turbine is considerably higher than that of the surroundings, the gas flow is so rapid that only a small quantity of heat is lost by each unit mass of gas, so we may take $q = 0$. Also there is usually no difference in elevation at each end. Hence our energy equation [3.16] becomes

$$w = h_1 - h_2 + 1/2\,(v_1^2 - v_2^2) \qquad [3.17]$$

We are thus able to calculate the work obtainable from a knowledge of the enthalpy and velocity difference of the gas entering and leaving the turbine.

Flow through a nozzle

When a gas flowing down a pipe encounters a change in the cross-sectional area, there is a change of gas velocity. We utilise this effect frequently in engineering and in particular in a turbine where we 'throw' the gas on to the turbine blades with a high velocity. The incoming gas (steam in the case of a steam turbine) is speeded up by passing it through a nozzle, as in Fig. 3.7. No shaft work w is done, the system is assumed to be horizontal and we further assume that no heat q enters the system as the gas flow is too rapid for this to be appreciable. Equation [3.16] then becomes:

$$v_1{}^2 - v_2{}^2 = 2(h_2 - h_1) \qquad [3.18]$$

which relates the velocity change to the enthalpy change.

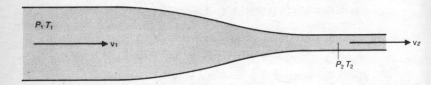

Fig. 3.7 A steady flow process through a nozzle.

In practice, we will be given the 'upstream' conditions P_1, T_1 which means that we know h_1. However, if as is usual, only the 'downstream' pressure P_2 is specified, we have insufficient information to determine h_2. If we now assume the flow through the nozzle to be reversible as well as adiabatic and the gas is treated as ideal, then we can find the downstream temperature T_2 from the adiabatic relation

$$\left(\frac{T_1}{T_2}\right)^{\gamma} = \left(\frac{P_1}{P_2}\right)^{\gamma - 1}$$

This gives us sufficient information to find h_2 and therefore v_2.

Chapter 4
The second law of
thermodynamics

The first law of thermodynamics tells us that, in any process, energy is conserved. It may be converted from one form to another but the total amount of energy is unchanged. The second law of thermodynamics imposes limits on the efficiency of processes which convert heat into work, such as steam or internal combustion engines. It will allow us to set up the thermodynamic temperature scale which is independent of the nature of the thermometric substance, and to define the concept of entropy, which we shall see is related to the microscopic disorder in the system.

Before we set up the second law we shall first discuss Carnot cycles, which are central to the discussion.

4.1 Carnot cycles

At the beginning of the nineteenth century, when steam engines were in their infancy, there was enormous interest in how their efficiency could be increased. An intellectual giant in this field was a French engineer, Carnot, who published in 1824 a powerful paper on how work could be produced from sources of heat. He knew that work could be obtained from an engine if there were heat sources at different temperatures – the boiler and the surrounding air in the case of a steam engine. He also knew that it was possible for heat to flow from a hot body to a cold body with no work being performed, the flow continuing until thermal equilibrium was attained. Carnot realised then that since any return to thermal equilibrium could be used to produce work, any return to equilibrium without the production of this work must be considered a loss. So any temperature difference may be utilised in the production of work or it may be wastefully dissipated in a spontaneous flow of heat. He concluded that, in an efficient engine, all transfers of heat should be between bodies of nearly equal temperature. With these ideas in mind, he designed an idealised engine which is of fundamental significance. The cycle for the Carnot engine is depicted in Fig. 4.1.

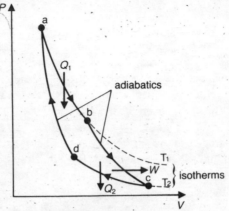

Fig. 4.1 A Carnot cycle for an ideal gas.

A working substance, which could be any substance but we have chosen it to be an ideal gas for the purpose of our discussion, is taken round the reversible cycle abcd. ab is an isotherm at the temperature T_1 and heat Q_1 enters from a heat reservoir at T_1. cd is an isotherm at a lower temperature T_2 where heat Q_2 is rejected to another reservoir at that temperature. bc and da are adiabatics. The work W done in the cycle is the area abcd.

It is important to emphasise that a Carnot engine operates between *only two reservoirs* and that it is *reversible*. Also, if a working substance is chosen other than an ideal gas, then the shape of the Carnot cycle as depicted in Fig. 4.1 is different because the equations for the adiabatics and isotherms are no longer $PV^\gamma = $ constant and $PV = nRT$.

It is interesting to note that Carnot's ideas were conceived before the first law was formulated, using the caloric concept of heat.

4.2 Efficiency of an engine

Any general heat engine E may be represented by the schematic diagram Fig. 4.2 where the heat *supplied* Q_1 and the heat *rejected* Q_2 are not necessarily obtained from just two heat reservoirs as in the special case of the Carnot engine. W is the work done *by* the engine. The arrow around the edge of the block depicting the engine indicates that the latter works in a cycle.

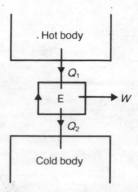

Fig. 4.2 A schematic representation of an engine working in a cycle. The efficiency $\eta = W/Q_1 = 1 - Q_2/Q_1$.

As a measure of 'what we get out for what we put in', we define the efficiency η of an engine as the work performed divided by the heat put in. For the engine cycle depicted in Fig. 4.2,

$$\eta = W/Q_1$$

Applying the first law to the working substance in the engine,

$$\Delta U = Q_1 - Q_2 - W$$

In writing this, we have remembered our sign convention where the heat *into* the system and the work done *on* the system are both counted positively. As the working substance is unchanged in a cycle, $\Delta U = 0$ and our first law becomes

$$W = Q_1 - Q_2$$

or
$$\eta = 1 - Q_2/Q_1 \qquad\qquad [4.1]$$

4.3 Statements of the second law of thermodynamics

There are two statements of the second law of thermodynamics which are both based on our general experience. They were each formulated in the eighteen-fifties by Clausius and Kelvin, but the second was subsequently modified by Planck. We shall shortly show that both statements are equivalent and they are as follows, although not in the original words.

The Kelvin–Planck statement

It is impossible to construct a device that, operating in a cycle, will produce no other effect than the extraction of heat from a single body at a uniform temperature and the performance of an equivalent amount of work.

Schematically this statement is represented in Fig. 4.3(a). The second law implies that some heat must also be rejected by the device to a body at a lower temperature; otherwise, as can be seen from equation [4.1], one could have an engine with 100 per cent efficiency. Were this statement untrue, we could drive a ship across the sea just by extracting heat from the sea and converting it totally into work!

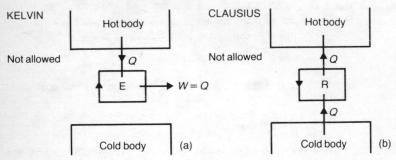

Fig. 4.3 Schematic representations of the Kelvin and Clausius statements of the second law of thermodynamics.

There are two key words in this statement which necessitate discussion.

1. *Cycle* requires that the working substance be *unchanged*. Many

processes can be thought of which convert heat completely into work, but in all of them the working system is also changed. For example, we could heat one mole of an ideal gas and allow it to expand quasistatically and isothermally (by keeping it in contact with a thermal reservoir) from a volume V_1 to V_2. The work done *by* the gas is

$$W = \int_{V_1}^{V_2} P\,\mathrm{d}V = RT \int_{V_1}^{V_2} \frac{\mathrm{d}V}{V} = RT\ln(V_2/V_1)$$

As the expansion is isothermal, $T_1 = T_2$ and so $\Delta U = 0$. The first law then shows that $Q = W$ where Q is the net heat supplied, and we have a 100 per cent conversion of heat into work. But there is no violation of the second law here as there has been a *change in the system*.

2. The other key word is *single*. Suppose that the heat $Q_1 + Q_2$ is supplied from two bodies: Q_1 from a body at T_1 and Q_2 from a body at T_2, with $T_1 > T_2$, say. The cyclical engine delivers an amount of work $W = Q_1 + Q_2$ as shown in Fig. 4.4.

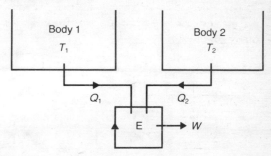

Fig. 4.4 In the Kelvin statement of the second law, heat has to be extracted from a *single* body.

However, such an engine does not violate the second law because Q_2 could be negative with $W = |Q_1| - |Q_2|$. We thus have to exclude this possible type of engine from the Kelvin statement by specifying a *single* body.

The Clausius statement

> *It is impossible to construct a device that, operating in a cycle, produces no other effect than the transfer of heat from a cooler to a hotter body.*

Schematically, this statement is represented in Fig. 4.3(b). This form of the second law tells us that work must be performed if heat is to be transferred from a colder to a hotter body. Were this not so, we could heat our houses just by cooling the ground, at no cost and with no work having to be done!

An engine which extracts heat from a cold body and delivers heat to a hot body when work is performed on the engine is called a *refrigerator*. (In our diagrams, we shall denote refrigerators by R.)

There is one final point that should be discussed. The Kelvin statement of the second law refers to the impossibility of heat being extracted from a hot body and the performance of an equivalent amount of work, with there being no change in the working system. It does *not* forbid the *opposite* situation, depicted in Fig. 4.5, where all the work W done on an unchanged system may be converted *completely* into heat. Rumford's experiment with a blunt boring tool is an example of such a total conversion of work into heat.

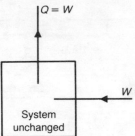

Fig. 4.5 Work can be converted completely into heat.

4.4 The equivalence of the Kelvin and Clausius statements

The two statements of the second law of thermodynamics may be shown to be equivalent by showing that the falsity of each implies the falsity of the other.

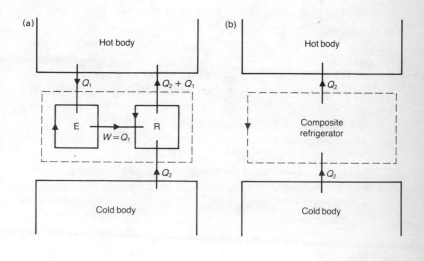

Fig. 4.6 If the Kelvin statement of the second law is false, this implies that the Clausius statement is also false. We use the arrangement illustrated here to prove this.

Let us suppose first that the Kelvin statement is untrue. This means that we can have an engine E which takes Q_1 from a hot body and delivers work $W = Q_1$ in one cycle. Let this engine drive a refrigerator R as shown in Fig. 4.6(a) and let us now adjust the size of the working cycles so that W is sufficient work to drive the refrigerator round one cycle. Suppose the refrigerator extracts heat Q_2 from the cold body. Then the heat delivered by it to the hot body is $Q_2 + W$ or $Q_1 + Q_2$. We may regard the engine and the refrigerator as the composite engine enclosed by the dotted line as shown in Fig. 4.6(b). This composite engine (strictly a refrigerator) extracts Q_2 from the cold body and delivers a net amount of heat $Q_2 + Q_1 - Q_1 = Q_2$ to the hot body, but no work is done. Hence we have a violation of the Clausius statement.

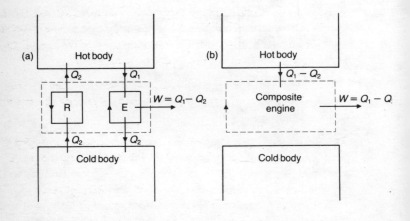

Fig. 4.7 If the Clausius statement of the second law is false, this implies that the Kelvin statement is also false. We use the arrangement illustrated here to prove this.

Suppose now that the Clausius statement is untrue. This means that we can have a refrigerator which extracts heat Q_2 from a cold body and delivers the same heat Q_2 to a hot body in one cycle, with no work having to be done. Let us now have an engine which operates between the same two bodies and let us adjust the size of its working cycle so that, in one cycle, it extracts heat Q_1 from the hot body, gives up the same heat Q_2 to the cold body as was extracted by the refrigerator and so delivers the work $W = Q_1 - Q_2$. This is depicted in Fig. 4.7(a). The engine and the refrigerator may be regarded as a composite engine enclosed by the dotted line, as shown in Fig. 4.7(b), which takes in heat $(Q_1 - Q_2)$ from the hot body and delivers the same amount of work. Hence, we have a violation of Kelvin. This proves the equivalence of the two statements.

4.5 Carnot's theorem

In the introduction to this chapter, we saw that Carnot had argued that efficient engines must be those operating as nearly as possible to a Carnot cycle. Using our Clausius statement of the second law, let us now prove Carnot's theorem which states:

> No engine operating between two reservoirs can be more efficient than a Carnot engine operating between those same two reservoirs.

To prove this, let us imagine that such a hypothetical engine E' does exist with an efficiency η'. As shown in Fig. 4.8(a), this engine extracts heat Q_1' from the hot reservoir, performs the work W' and delivers the heat $Q_2' = (Q_1' - W')$ to the cold reservoir.

Let us now operate a Carnot engine, denoted by C and with efficiency η_C, between the two reservoirs extracting and delivering the heats Q_1 and Q_2, and let us also adjust the size of the cycle to make this engine perform the same amount of work as the hypothetical engine. For this Carnot engine, $Q_2 = Q_1 - W$. As the hypothetical engine is assumed to be more efficient than the Carnot engine,

$$W'/Q_1' > W/Q_1 \quad (W' = W)$$

so $\quad Q_1 > Q_1'$

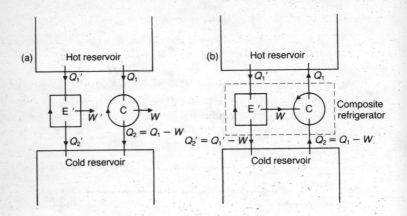

Fig. 4.8 The arrangement used to prove Carnot's theorem: no engine working between two reservoirs can be more efficient than a Carnot engine working between the same two reservoirs.

Now a Carnot engine is a reversible engine so we may drive it backwards as a refrigerator as shown in Fig. 4.8(b). The hypothetical engine and the Carnot refrigerator together act as a composite device, shown by the dotted line, which extracts *positive* heat $(Q_1 - Q_1')$ from the cold reservoir and delivers the *same* heat to the hot reservoir with no external work being required. But reservoirs are just large bodies where the temperature is unchanged upon the addition of heat. This means that we have a violation of the Clausius formulation and so the engine E' cannot exist and our original assumption that $\eta' > \eta_C$ is incorrect.

We conclude that, for any real engine,

$$\boxed{\eta \leqslant \eta_C}$$

which proves the theorem.

4.6 Corollary to Carnot's theorem

It follows from Carnot's theorem that:

> *All Carnot engines operating between the same two reservoirs have the same efficiency.*

To prove this statement, let us imagine two Carnot engines C and C' operating between the same two reservoirs, and let the size of the working cycles be adjusted so that they each deliver the same amount of work.

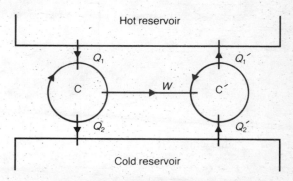

Fig. 4.9 The arrangement used to prove that all Carnot engines operating between the same two reservoirs have the same efficiency.

Let C run C' backwards as in Fig. 4.9. It follows from the argument just given in the previous section that

$$\eta_C \leqslant \eta_C{}'$$

If C' now runs C backwards,

$$\eta_C{}' \leqslant \eta_C$$

We conclude that

$$\boxed{\eta_C = \eta_C{}'}$$

which proves our assertion.

4.7 The thermodynamic temperature scale

We have just seen that the efficiency of a Carnot engine operating between the two reservoirs is independent of the nature of the working substance and so can be dependent only on the temperatures of the reservoirs. This gives us a means of defining a temperature scale which is independent of any particular material. Let us define the thermodynamic temperature T so that T_1 and T_2 for the two reservoirs in a Carnot engine are related as

$$\eta_C = \frac{T_1 - T_2}{T_1} = 1 - T_2/T_1 \qquad [4.2]$$

If we compare this with equation [4.1], we have

$$\boxed{T_1/T_2 = Q_1/Q_2} \qquad \text{(Carnot)} \qquad [4.3]$$

where we have written Carnot in brackets to remind us that this definition holds only for a Carnot engine.

We can see that equation [4.3] gives a sensible definition for a scale of temperature by considering Fig. 4.10. Here, we have a Carnot engine C_{12} operating between the reservoirs at T_1 and T_2. For this engine, equation [4.3] gives

$$T_1/T_2 = Q_1/Q_2 \qquad [4.4]$$

Suppose we have a second Carnot engine C_{23} operating between the reservoir at T_2 and a third reservoir at T_3. Let C_{23} absorb the same amount of heat Q_2 from the reservoir at T_2 as was rejected to that reservoir by C_{12}. When the two engines operate together, the reservoir at T_2 is thus unchanged. Equation [4.3] gives

$$T_2/T_3 = Q_2/Q_3 \qquad [4.5]$$

Multiplying equation [4.4] by equation [4.5],

$$T_1/T_3 = Q_1/Q_3$$

which does not involve the intermediate temperature T_2. As the reservoir at T_2 is unchanged, we may consider the two engines C_{12} and C_{23}, acting together, to be a composite Carnot engine C_{13} operating between the two reservoirs at T_1 and T_3. This composite engine is denoted by the dotted line in Fig. 4.10. The application of equation [4.3] again shows that the previous relation is precisely the one that holds for this composite Carnot engine. It follows that, by taking a whole series of Carnot engines, any range of temperatures may be defined in a self-consistent way.

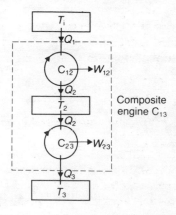

Fig. 4.10 The thermodynamic temperature scale, as defined by equation [4.3], is consistent with the arrangement illustrated here.

This temperature scale is *independent of the choice of working substance*, which was one of our objectives in our discussion of scales of temperature in Chapter 1. The thermodynamic scale of temperature will now be shown to be identical to the familiar ideal gas scale.

4.8 The equivalence of the thermodynamic and the ideal gas scales

Until now, we have used the symbol T for absolute temperature as defined on the ideal gas scale. In this section, until we prove them to be identical, we shall use the symbol T_g for the gas scale temperature and T for the thermodynamic temperature as just defined.

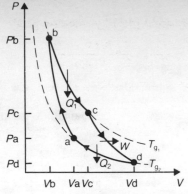

Fig. 4.11 A Carnot cycle with an ideal gas as the working substance. Using this figure, we show in the text that the ideal gas scale temperature is identical to the thermodynamic temperature T.

Consider a Carnot engine, with an ideal gas as the working substance, operating between the two reservoirs at the ideal gas scale temperatures T_{g_1} and T_{g_2}. Let us follow the operating cycle abcd shown in Fig. 4.11. For the isotherm bc, we have the empirical equation of state involving the gas scale temperature T_{g_1}

$$PV = nRT_{g_1} \qquad [4.6]$$

The first law gives for an infinitesimal part of this reversible process

$$\dbar Q = dU + PdV = PdV \qquad [4.7]$$

as $dU = 0$ because the temperature is constant. The heat Q_1 entering the engine in this portion of the cycle is

$$Q_1 = \int_{V_b}^{V_c} PdV = nRT_{g_1} \int_{V_b}^{V_c} dV/V$$

$$= nRT_{g_1} \ln (V_c/V_b) \qquad [4.8]$$

This is positive if $V_c > V_b$, which is consistent with our idea that heat enters the engine in this portion of the cycle. Similarly, the heat *entering* the engine along the da isotherm part of the cycle is $nRT_{g_2} \ln (V_a/V_d)$. This is negative if $V_a < V_d$, which means that heat flows *out* of the engine. But we have defined positive Q_2 as the heat flow *out* of the engine here, so

$$Q_2 = - nRT_{g_2} \ln (V_a/V_d) = nRT_{g_2} \ln (V_d/V_a) \qquad [4.9]$$

Dividing equation [4.8] by equation [4.9],

$$\frac{Q_1}{Q_2} = \frac{T_1}{T_2} = \frac{T_{g_1} \ln (V_c/V_b)}{T_{g_2} \ln (V_d/V_a)} \qquad [4.10]$$

But ab and cd are adiabatics where $T_g V^{\gamma - 1}$ = a constant holds, so:

$$T_{g_1} V_c^{\gamma - 1} = T_{g_2} V_d^{\gamma - 1} \qquad \text{(cd adiabatic)} \quad [4.11]$$

$$T_{g_1} V_b^{\gamma - 1} = T_{g_2} V_a^{\gamma - 1} \qquad \text{(ab adiabatic)} \quad [4.12]$$

Dividing equation [4.11] by equation [4.12],

$$\frac{V_c}{V_b} = \frac{V_d}{V_a}$$

or $\quad \ln \left\{ \dfrac{V_c}{V_b} \right\} = \ln \left\{ \dfrac{V_d}{V_a} \right\}$

Substituting this in equation [4.10],

$$\frac{T_1}{T_2} = \frac{T_{g_1}}{T_{g_2}}$$

This means that

$$T_g = \epsilon T$$

where ϵ is a constant. But we know that all temperature scales agree at the fixed point of 273.16 K so ϵ must be unity. We conclude then that

$$\boxed{T_g \equiv T} \qquad [4.13]$$

That is, *the thermodynamic and the ideal gas scales of temperature are identical.*

4.9 The efficiencies of engines and refrigerators using Carnot cycles

In the next section we shall consider an example of a real engine, but it is instructive to consider first, because the analysis is so delightfully simple, an engine based on a Carnot cycle. This has a practical use in that it gives us an upper limit, by Carnot's theorem, for the efficiency of any possible engine that we might design.

A Carnot engine, such as the one depicted in Fig. 4.1, has the efficiency

$$\eta_C = 1 - Q_2/Q_1 = 1 - T_2/T_1 \qquad [4.2]$$

It is then a simple matter to calculate the efficiency knowing T_1 and T_2. It is interesting to note that the efficiency would be 100 per cent were we able to obtain a lower temperature reservoir at absolute zero; this we shall see is forbidden by the third law.

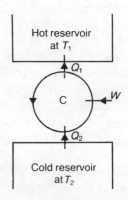

Fig. 4.12 A Carnot refrigerator.

Imagine now that the Carnot engine is run backwards, as in Fig. 4.12, to act as a refrigerator. We can define the 'efficiency' $\eta_C{}^R$ of a refrigerator as the heat extracted divided by the work expended and this is customarily known as the *coefficient of performance*. For our Carnot refrigerator,

$$\eta_C{}^R = \frac{Q_2}{W} = \frac{Q_2}{Q_1 - Q_2} = \frac{T_2}{T_1 - T_2} \qquad [4.14]$$

If we take the realistic values of $T_1 = 293$ K and $T_2 = 273$ K, $\eta_C{}^R = 13.6$. Real refrigerators have coefficients of performance of about 4 or 5.

The most interesting application of these ideas, and at first sight a rather disquieting one, is to so-called *heat pumps*. We all know that the back of a refrigerator becomes rather warm. Let us define, as a measure of how good a refrigerator is at providing heat, the 'efficiency' η^{HP} of a heat pump to be the quantity Q_1/W. For a Carnot heat pump

$$\eta_C{}^{HP} = \frac{Q_1}{W} = \frac{Q_1}{Q_1 - Q_2} = \frac{T_1}{T_1 - T_2} = \frac{1}{1 - T_2/T_1} \quad [4.15]$$

Suppose now that $T_1 = 293$ K and $T_2 = 273$ K. Then $\eta_C{}^{HP} = 15$, so we obtain 15 joules of heat for every joule of work we put in! Of course this is an idealised and most efficient device but real heat pumps have been produced for some forty years and, although their efficiencies are not 15, they are in the region of 3 or 4, which makes them attractive propositions. Their disadvantage is their high capital cost. It is expensive to lay pipes under one's garden in order to cool this to heat one's house. The Festival Hall in London is heated by operating a heat pump between it and the Thames, and one Oxford college derived, until recently, its warmth by cooling the city sewers.

Fig. 4.13 shows the efficiency of a Carnot heat pump against (T_2/T_1). Because the efficiency rises as the temperature difference between the reservoirs decreases, heat pumps are best used in providing background heating, with the final 'top up' being provided by a conventional source.

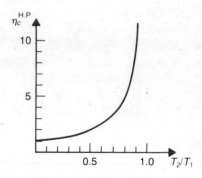

Fig. 4.13 The 'efficiency' of a Carnot heat pump as a function of the ratio of the reservoir temperatures.

4.10 Real engines

The Carnot engine is an idealised engine. Real engines operate in various cycles, all different from the idealised Carnot one. At this stage we shall temporarily depart from the central thread of our argument to discuss, for interest, the four-stroke Otto-cycle or common petrol engine.

We are all familiar with the principle of this engine.

1. Petroleum vapour and air are drawn into the cylinder on the downstroke of the piston.
2. The mixture is compressed.
3. Near top dead centre, the mixture is ignited giving rise to the power stroke.
4. The burnt mixture is expelled.

Fig. 4.14(a) shows this cycle. The air standard Otto cycle is a close but simplified representation of this process which facilitates analysis; it is shown in Fig. 4.14(b). The working substance is assumed to be just air rather than air and petrol, with no chemical changes occurring in its composition. Also, instead of the heat being added internally by the burning of the fuel, heat is assumed to be added from external sources. The suction intake and the exhaust processes of the actual cycle, shown in Fig. 4.14(a), are omitted from the Otto cycle.

Let us now go round the Otto cycle.

a–b The piston rises to compress the gas reversibly and adiabatically with

$$T_a V_1^{\gamma - 1} = T_b V_2^{\gamma - 1} \qquad [4.16]$$

b–c Heat Q_1 is added at constant volume from an external source with

$$Q_1 = C_V (T_c - T_b) \qquad [4.17]$$

c–d The gas expands adiabatically and reversibly in the power stroke with

$$T_d V_1^{\gamma - 1} = T_c V_2^{\gamma - 1} \qquad [4.18]$$

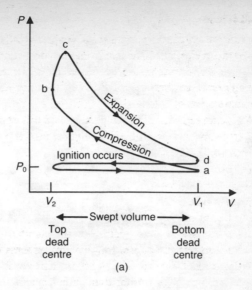

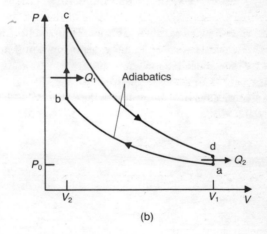

Fig. 4.14 The 4-stroke internal combustion engine. (a) The actual cycle.
(b) The Otto cycle, which is a simplified representation of the actual cycle.

d–a At the bottom of the power stroke the gas is assumed to cool
at constant volume down to the pressure P_0 by giving up heat
Q_2 to external reservoirs with

$$Q_2 = C_V(T_d - T_a) \qquad \text{[4.19]}$$

We may now derive an expression for the efficiency using equations [4.1], [4.17] and [4.19]. For this cycle,

$$\eta = 1 - Q_2/Q_1 = 1 - \left\{ \frac{T_d - T_a}{T_c - T_b} \right\}$$

The two relations, equations [4.16] and [4.18], for the adiabatic processes, give no subtraction

$$(T_d - T_a) V_1^{\gamma - 1} = (T_c - T_b) V_2^{\gamma - 1}$$

or
$$\left(\frac{V_1}{V_2} \right)^{\gamma - 1} = \left\{ \frac{T_c - T_b}{T_d - T_a} \right\}$$

We call the ratio V_1/V_2 the compression ratio r_c, so

$$\eta = 1 - \left(\frac{V_2}{V_1} \right)^{\gamma - 1} = 1 - \frac{1}{r_c^{\gamma - 1}} \qquad \text{[4.20]}$$

and it can now be seen why it is important to have as high a compression ratio as possible. Pinking or pre-ignition limits r_c to about 7 or 8 for unleaded low-octane petrol, giving a theoretical efficiency of $1 - 1/7^{0.4} = 54$ per cent. The actual efficiency of a real petrol engine is much lower than this value, being probably only about 30 per cent.

4.11 Summary

This has been a very important chapter and it is useful to summarise what we have found.

1. A heat engine converts heat into work in a cyclical process in which the working substance is unchanged.
2. A Carnot engine is a reversible engine which operates between two temperatures only. In general, engines take in and reject heat at a variety of temperatures.
3. The efficiency of an engine is

$$\eta = 1 - Q_2/Q_1$$

4. The essence of the Kelvin statement of the second law is that a cyclical engine cannot convert heat from a single body at a uniform temperature completely into work. Some heat has to be rejected at a lower temperature.

   The essence of the Clausius statement is that heat cannot flow from a cold body to a hot body by itself – work has to be done on a cyclical refrigerator to achieve this.

5. The most efficient engine operating between a given pair of reservoirs is a Carnot engine. All Carnot engines operating between the same reservoirs have the same efficiency, this being independent of the nature of the working substance.

6. For a Carnot engine, we can define the thermodynamic temperature as

$$\frac{Q_1}{Q_2} = \frac{T_1}{T_2}. \quad \text{with} \quad \eta_C = 1 - T_2/T_1$$

The thermodynamic temperature is identical to the ideal gas temperature.

We are now in a position to meet the powerful concept of entropy.

Chapter 5
Entropy

5.1 The Clausius inequality

There is a very important theorem for cyclical processes which leads to the concept of entropy. This theorem is known as the Clausius inequality.

Let us consider a working substance undergoing a cycle so that, at the end of the cycle, its state is unchanged. In Fig. 5.1(a) we symbolically represent this cycle by the circle in the centre with the starting state at the temperature T_1 being represented by the point 1. We shall take the heat causing the changes as ultimately being supplied by a principal reservoir at $\widetilde{T}$. We can take the working substance around the cycle in the following way.

The state of the working substance is first changed to an infinitesimally close neighbouring state 2 at a temperature T_2 by injecting a small amount of heat δQ_1. We do this with a Carnot engine C_1, which operates between two auxiliary reservoirs at T_1 and $\widetilde{T}$. The auxiliary reservoir at T_1 supplies δQ_1 to the working substance and an equal quantity of heat is supplied by C_1 to that reservoir to leave it unchanged. C_1 in its turn takes heat $\{\widetilde{T}/T_1\}\delta Q_1$ from the auxiliary reservoir at $\widetilde{T}$ and performs work, δW_1 say. If the auxiliary reservoir at $\widetilde{T}$ is to remain unchanged, heat $\{\widetilde{T}/T_1\}\,\delta Q_1$ enters it from the principal reservoir. In this way, we can effect a change from 1 to 2 with the only other changes being the performance of the external work δW_1 and the extraction of heat $\{\widetilde{T}/T_1\}\,\delta Q_1$ from the principal reservoir.

The process is repeated taking the working substance from 2 to 3 with the help of the Carnot engine C_2 and a new pair of auxiliary reservoirs at T_2 and $\widetilde{T}$, and so on round the cycle.

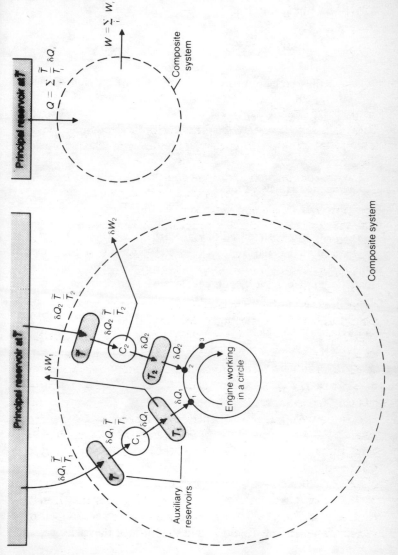

Fig. 5.1 The cycle used to derive the Clausius inequality.

Consider now the composite system consisting of the working system, all the Carnot engines and all the auxiliary reservoirs. This composite system includes everything within the dashed line in Fig. 5.1 (a). At the end of the cycle:

1. everything in the composite system is unchanged and so $\Delta U = 0$;
2. the heat supplied to it is

$$Q = \sum_i \delta Q_i \frac{T}{T_i}$$

where the summation is over the number of different Carnot engines used;

3. the external work performed is

$$\sum_i \delta W_i = W, \text{ say}$$

Applying the first law to the composite system,

$$0 = Q - W \quad \text{or} \quad W = Q$$

This situation is represented in Fig. 5.1(b), where we can see that we have extracted heat from a *single* reservoir and have performed an equal amount of work. This is a violation of the Kelvin statement of the second law. The only way this process can occur is for both W and Q to be negative, that is, work is done *on* the system and an equal quantity of heat flows *out*. This is just the allowed situation of Fig. 4.5. Alternatively, both W and Q could be zero.

We conclude that

$$W = Q \leqslant 0$$

i.e. $\quad T \sum_i \frac{\delta Q_i}{T_i} \leqslant 0 \quad$ or $\quad \sum_i \frac{\delta Q_i}{T_i} \leqslant 0$

In the limit,

$$\oint \frac{đQ}{T} \leqslant 0$$

where the circle on the integral sign indicates that the cycle is complete or closed. This is known as the *Clausius inequality* and is one of the key results in thermodynamics.

There are three important points that should be made.

1. The proof of the inequality emphasises that the T appearing inside the integral is the temperature of the auxiliary reservoirs supplying heat to the working substance. It is thus the *temperature of the external source of heat*. We shall always write the Clausius inequality as

$$\oint \frac{đQ}{T_0} \leqslant 0 \qquad \text{(Clausius inequality)} \qquad [5.1]$$

where we have written T_0 to remind us of this.

2. If the cycle is reversible, the cycle could be undertaken in the opposite direction and our proof would give

$$\oint \frac{đQ}{T_0} \geqslant 0$$

(W would then be done *on* the composite system, with an equal amount of heat $\tilde{T} \sum_i \delta Q_i / T_i$ being rejected to the principal reservoir. This will not violate the Kelvin statement if $W = Q = \tilde{T} \sum_i \delta Q_i / T_i \geqslant 0$.)

The only way for both inequalities to be satisfied is for

$$\oint_R \frac{đQ_R}{T} = 0 \qquad \text{(reversible cycle only)} \qquad [5.2]$$

We have added R to the bottom of the integral sign and as a subscript to $đQ$ to remind us that this relation is valid only for a reversible process. However, we have dropped the 0 subscript on T as there is now no difference between the temperature of the external source supplying the heat and the temperature of the working substance.

3. One can never forget the sign of the inequality if one remembers that, in the proof, heat was always flowing into the engine so $T_0 > T$ and that the equality sign holds for the reversible case where $T_0 = T$. Replacing T by the larger T_0 makes the inequality less than zero.

5.2 Entropy

This concept follows immediately from the previous section. Suppose we were to take a system along a reversible path R_1 from an initial state i to a final state f and then back again to the initial state along another reversible path R_2, completing a reversible cycle. In Fig. 5.2 we have illustrated this for a gas system.

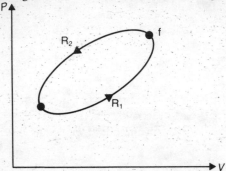

Fig. 5.2 A reversible cycle. We show in the text that $\int dQ/T$ is the same for the reversible paths R_1 and R_2 connecting i and f.

As the cycle is reversible, the equality sign holds in the Clausius inequality. Remembering that the cycle is composed of the two reversible paths, R_1 and R_2, we have

$$\oint_R \frac{dQ_R}{T} = \int_{R_1 i}^{f} \frac{dQ_R}{T} + \int_{R_2 f}^{i} \frac{dQ_R}{T} = 0$$

so

$$\int_{R_1 i}^{f} \frac{dQ_R}{T} = - \int_{R_2 f}^{i} \frac{dQ_R}{T}$$

But

$$\int_{R_2 i}^{f} \frac{dQ_R}{T} = - \int_{R_2 f}^{i} \frac{dQ_R}{T}$$

as R_2 is reversible. Thus,

$$\int_{R_1 i}^{f} \frac{dQ_R}{T} = \int_{R_2 i}^{f} \frac{dQ_R}{T}$$

which means that the integral $\int_{R i}^{f} dQ_R/T$ is *path independent*. We conclude that there must be a state function S with

$$\Delta S = S_f - S_i = \int_{\underset{R}{i}}^{f} \frac{\text{d}Q_R}{T} \qquad [5.3]$$

We call this state function the *entropy*. Notice that only entropy differences have been defined. Also, it cannot be stressed too strongly that *the defining integral for entropy differences has to be taken over a reversible path*.

For an infinitesimal reversible process,

$$\text{d}Q_R = T\,\text{d}S \qquad \text{(reversible only)} \quad [5.4]$$

The name entropy comes from the Greek *en* meaning 'in' and *trope* meaning 'turning'. Clausius intended the word to convey the idea of heat being converted in an engine into work.

5.3 An example of a calculation of an entropy change

So that we may see how entropy changes are calculated, let us determine the entropy change of a beaker of water when it is heated at atmospheric pressure between room temperature, at 20°C, and 100 °C by placing it on a reservoir at 100 °C. When the water reaches 100 °C, the beaker is removed from the reservoir and placed in an insulating jacket. This process is shown in Fig. 5.3. Heat passes from the reservoir into the water and it might seem that a simple application of equation [5.3] would suffice. However this equation applies to a *reversible* process, while the actual process here is *irreversible* because of the inherent finite temperature differences.

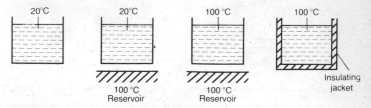

Fig. 5.3 A beaker of water is heated irreversibly and isobarically between 20 °C and 100 °C in this process.

We resort again to the argument encountered in section 2.4. As the water is initially and finally in equilibrium states, with well-defined entropies, the entropy change for this process is also well defined. We can then *imagine any convenient reversible process* that takes the system between the same two end points and calculate, using equation [5.3], the entropy change for this imaginary process. *This entropy change is then the same as that occurring in the actual irreversible process.*

One simple reversible heating process between the end points could be effected by bringing up a whole series of reservoirs between 20 °C and 100 °C, keeping the pressure constant, so that the water passes through a series of equilibrium states. This imaginary process is shown in Fig. 5.4.

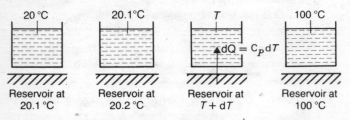

| 20 °C | 20.1°C | T | 100 °C |

$$dQ = C_P dT$$

| Reservoir at 20.1 °C | Reservoir at 20.2 °C | Reservoir at $T + dT$ | Reservoir at 100 °C |

Fig. 5.4 The same beaker of water is heated reversibly and isobarically between 20 °C and 100 °C in this imaginary process.

When the water is at T and it is heated to $T + dT$ by the reservoir at $T + dT$, the heat entering the water reversibly is

$$đQ_R = C_P \, dT$$

where C_P is the heat capacity at constant pressure of the water. Hence the entropy change of the water is

$$dS = \frac{C_P dT}{T}$$

or $$\Delta S = C_P \int_{T_i}^{T_f} \frac{dT}{T} = C_P \ln \left\{ \frac{T_f}{T_i} \right\} \qquad [5.5]$$

If we have 1 kg of water with $C_P = 4.2$ kJ K^{-1} and remembering that the T in equation [5.3] is the thermodynamic temperature,

$$\Delta S = 4.2 \times 10^3 \ln \left\{ \frac{373}{293} \right\} = 1.01 \times 10^3 \text{ J K}^{-1}$$

Any other reversible path would give, of course, the same answer, but this path is probably the most convenient.

5.4 The entropy change in a free expansion

As a second example of a calculation of an entropy change, let us evaluate the entropy change of an ideal gas undergoing a free expansion doubling its volume (see sections 2.6, 3.6).

We know that in a free expansion: the process is irreversible; there is no temperature change; and no heat enters the system because the walls are adiabatic. In order to apply equation [5.3] we imagine a reversible isothermal doubling of the volume and calculate the entropy change for this. Such an expansion could be achieved by allowing the gas to expand slowly but being in thermal contact with a reservoir at T. Applying the first law to this process,

$$đQ = dU + P\,dV = P\,dV$$

as $dU = 0$ because T is constant. Thus

$$dS = \frac{P}{T}\,dV = \frac{nR}{V}\,dV \quad \text{as} \quad PV = nRT$$

So $\quad \Delta S = nR \int_{V}^{2V} \frac{dV}{V} = nR \ln 2$ [5.6]

and this is the entropy change in the actual irreversible free expansion.

It is often mistakenly thought that heat has to flow into a system for there to be an entropy change and, conversely, that *any* adiabatic process takes place at constant entropy, or isentropically. This example shows this not to be so. As

$$đQ_R = T\,dS$$

applies to a reversible process only, a process has to be *both* adiabatic and reversible to be isentropic. Although our free expansion is adiabatic, it is not isentropic because it is irreversible.

5.5 The principle of increasing entropy

There is contained in the Clausius inequality the profound implication that processes can occur only if the net entropy of the universe increases or stays the same. To see how this arises, let us consider the cycle shown in Fig. 5.5, consisting of

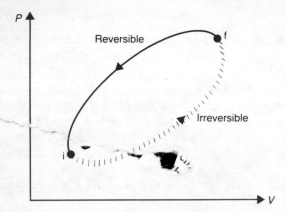

Fig. 5.5 An irreversible cycle consisting of an irreversible process followed by a reversible process back to the initial state again.

an irreversible 'path' i to f followed by a reversible path f back to i again. To be specific, we have chosen a gas system but the argument is general. The Clausius inequality gives

$$\oint \frac{dQ}{T_0} \leqslant 0$$

where the equals sign applies if the path i to f is reversible and so the whole cycle is reversible.

It follows that

$$\int_i^f \frac{dQ}{T_0} + \int_{R\,f}^i \frac{dQ_R}{T} \leqslant 0$$

or $$\int_i^f \frac{dQ}{T_0} \leqslant -\int_{R\,f}^i \frac{dQ_R}{T} = +\int_{R\,i}^f \frac{dQ_R}{T} = S_f - S_i$$

as the path $f \rightarrow i$ is reversible.

For an infinitesimal part of the process,

$$\boxed{\frac{\text{đ}Q}{T_0} \leqslant \text{d}S}$$ [5.7]

where the equality sign holds if it is reversible and then $T = T_0$. What this equation is saying is that, in an infinitesimal irreversible process between a pair of equilibrium states, there is a definite entropy change $\text{d}S$, but this is larger than the heat supplied in that irreversible process divided by the temperature of the external heat source. One must not confuse this heat with the heat supplied in any imaginary reversible process used to calculate $\text{d}S$.

Suppose now that the system is thermally isolated. Then,

$$\text{đ}Q = 0$$

and $\text{d}S \geqslant 0$

or $$\boxed{S_f - S_i = \Delta S \geqslant 0}$$ (thermally isolated) [5.8]

for a finite process. We conclude that:

The entropy of a thermally isolated system increases in any irreversible process and is unaltered in a reversible process. This is the principle of increasing entropy.

If in addition to being thermally isolated, the system is mechanically isolated from the surroundings so that no work can be done, by the first law, the internal energy U remains constant too for this condition of total isolation.

One word of warning must be given here. This statement refers to *net* entropy changes. It does not say that the entropy of part of the system cannot go down. In Fig. 5.6, for example, heat flows from body A to body B at a lower temperature, both of which are contained in an adiabatic enclosure. ΔS^A is then negative but

$$\Delta S = \Delta S^A + \Delta S^B$$

will still, by the entropy increasing principle, be positive.

Before proceeding further, we must be absolutely clear as to the meaning of the entropies S_i and S_f in equation [5.8]. We have

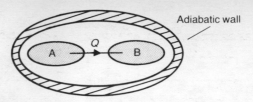

Fig. 5.6 Although the entropy of a thermally isolated system can only increase, or remain the same, the entropy of part of the system can decrease.

been considering an adiabatic process in which the system is changed from some initial equilibrium state with an entropy S_i to a final equilibrium state with an entropy S_f. This entropy change could be brought about by a variety of means. For example, work could be performed on the system irreversibly, or the system could consist of two parts, A and B, which are initially at different temperatures and thermally insulated from one another; we then allow heat to flow by removing the insulator, as indicated in Fig. 5.7.

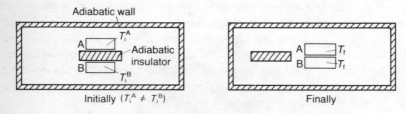

Fig. 5.7 Entropy changes are calculated between equilibrium states.

Because we are talking about equilibrium states initially and finally, the concept of the entropy change should cause us no difficulty. However, consider now as our system a bar, in our adiabatic enclosure, with one end initially hotter than the other, as in Fig. 5.8(a). The hot end of the bar will cool and the cold end will warm so that the initial temperature gradient disappears and we are left with a state of uniform temperature. Is there an entropy change and can we apply our result equation [5.8]? The answer is yes, but we have to define the entropy of the initial non-equilibrium state in the following way. We imagine cutting the bar up into thin slices, which are then insulated from each

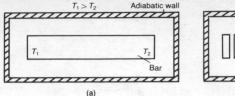

$T_1 > T_2$ Adiabatic wall

T_1 T_2
Bar

(a) (b)

Fig. 5.8 The entropy change of a bar in which there is an initial temperature gradient can be calculated by dividing the bar up into slices and considering the entropy change for each slice.

other, as in Fig. 5.8(b). The temperature of each slice may be
taken as being uniform over its thickness. Each slice may be
regarded as being in an equilibrium state with a particular value of
entropy determined by its mean temperature and the external
pressure. (We are making the physically reasonable assumption
that the entropy of a slice does not depend on the temperature
gradient, only on the mean temperature.) The entropy of the
whole bar may then be taken as the sum of the entropies of these
slices, and this idea can be used throughout the process so that we
can speak of the entropies of all the intermediate non-equilibrium
states for the bar. Interested readers are referred to the book by
Zemansky where the actual entropy change is calculated for such
a bar undergoing cooling (see Appendix 6).

Now this concept is no different from our initial ideas for our
simple system of Fig. 5.7. There, we regarded the initial entropy
of the system as the sum of the entropies for the two parts, considered as being isolated from each other. We then allowed them
to interact with heat flowing between them until we reached the
final state. At any intermediate non-equilibrium state we can still
think of the entropies of the two bodies just by thermally isolating them from each other.

We thus conclude that, even in a process starting from an initial
non-equilibrium state, the entropy of a thermally isolated system
increases until, as equilibrium is approached and the equality sign
of equation [5.8] has to be taken, it increases no more and so the
entropy reaches a maximum. In summary:

For a system thermally isolated from the surroundings:

    S → a maximum

For a system which is totally isolated from the surroundings:

    S → a maximum with U remaining constant

Finally, we can extend our argument by considering a system which is *not* thermally isolated but may exchange heat during a process with a *given* set of reservoirs. These reservoirs may also exchange heat amongst themselves but not with any others. Together with the original system, they form a combined system. Let us now surround this combined system with an adiabatic wall; this will not cause any other physical changes because no heat crosses this boundary. This adiabatic enclosure contains everything that interacts during the process under consideration and we may define this assembly as constituting our *universe* for the purpose of our thermodynamic argument. It should not be confused with the real universe, which may or may not be infinite and may or may not form an isolated system. Now, as our universe is thermally isolated,

$$\boxed{\Delta S^{\text{universe}} \geqslant 0}$$

                                                                    [5.9]

for a finite process, with the equality sign holding for a reversible process.

The fact that the entropy of a thermally isolated system can never decrease in a process provides a direction for the sequence of natural events. Newton's laws are second order in time t and are unaltered by replacing t with $-t$. This would seem to suggest that all physical processes can run backwards as well as forwards. Clearly this is not so: a teacup smashes into many pieces but we have never seen the pieces spontaneously reform again into the teacup. The law of increasing entropy tells us that processes go only in the direction of increasing entropy of the universe. In our example, the broken pieces have a higher entropy than the unbroken cup. It is for this reason that the law of increasing entropy is often described as providing 'the arrow of time' for the evolution of natural processes.

5.6 An example of a calculation of a net entropy change for the universe

We return to our previous example of heating a beaker of water, with thermal capacity C_P, from $T_i = 293$ K to $T_f = 373$ K. Let us calculate the net entropy change occurring. We have shown (equation [5.5]) that the entropy change of the water is

$$\Delta S^{\text{water}} = C_P \ln \left\{ \frac{T_f}{T_i} \right\} = C_P \ln \left(\frac{373}{293} \right)$$

What then is the entropy change, $\Delta S^{\text{reservoir}}$, of the reservoir at 100 °C? In this process the reservoir loses an amount of heat $Q = C_P (T_f - T_i)$ irreversibly. To calculate its entropy change, imagine the reservoir losing this heat reversibly. An imaginary way of achieving this is to bring up another reservoir at a slightly lower temperature and for this heat to be transferred. Then,

$$\Delta S^{\text{reservoir}} = \int \frac{\mathrm{d}Q_R}{T} = \frac{1}{T_f} \int \mathrm{d}Q_R = -C_P \frac{(T_f - T_i)}{T_f}$$

$$= -\frac{80\, C_P}{373}$$

as the temperature of the reservoir is constant at $T_f = 373$ K and the entropy change is negative as heat flows out. Thus

$$\Delta S^{\text{universe}} = C_P \left\{ \ln \left(\frac{373}{293} \right) - \frac{80}{373} \right\} = 0.027\, C_P$$

which is positive, as it should be for this irreversible process.

It is very instructive to modify this problem to ask what is $\Delta S^{\text{universe}}$ if the water is heated in two stages by placing it first on a reservoir at 50 °C and, when it has reached that temperature, transferring it to a second reservoir at 100 °C for the final heating?

As the water is still being taken between the same two states,

$$\Delta S^{\text{water}} = C_P \ln \left(\frac{373}{293} \right)$$

The net entropy change of the reservoirs can be found, using the same method that we have just employed, to be

$$\Delta S^{\text{reservoirs}} = -C_P \left\{ \frac{30}{323} + \frac{50}{373} \right\}$$

Hence

$$\Delta S^{\text{universe}} = C_P \left\{ \ln \left(\frac{373}{293} \right) - \frac{30}{323} - \frac{50}{373} \right\} = 0.014\, C_P$$

This is positive again, but less than the entropy change occurring when a single reservoir was employed. This is reasonable as the use of two reservoirs is *closer to a reversible heating*, employing a large number of reservoirs, than the use of just one.

Suppose, finally, that a Carnot engine is operated between the reservoir at 100 °C and the water, as in Fig. 5.9. If the operating

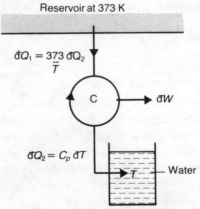

Fig. 5.9 A beaker of water may be heated reversibly by operating a Carnot engine between it and a reservoir at a higher temperature.

cycle of the engine is small so that the heat $ᵭQ_2$ rejected by the engine during one cycle causes only an infinitesimally small change dT in the temperature T of the water, T does not change significantly during one cycle and we have the required operating conditions for a Carnot engine of operating between a pair of reservoirs. As this actual process is reversible, we may employ equation [5.3] directly:

$$\Delta S^{\text{reservoir}} = -\int \frac{ᵭQ_1}{T^{\text{reservoir}}} = -\int \frac{ᵭQ_1}{373}$$

where dQ_1 is the heat given out by the reservoir in once cycle. But

$$\frac{dQ_1}{dQ_2} = \frac{373}{T}$$

Thus

$$dQ_1 = dQ_2 \frac{373}{T} = C_P \, dT \frac{373}{T}$$

and hence

$$\Delta S^{\text{reservoir}} = -\int_{293}^{373} \frac{373 \, C_P \, dT}{373 \, T} = -C_P \ln \frac{373}{293}$$

This is the negative of the entropy change of the water so

$$\Delta S^{\text{universe}} = 0$$

as it should be for this reversible process.

5.7 Entropy–temperature diagrams

The thermodynamic state of a system can be specified by any pair of independent state functions. In particular, a state is equally well specified by the pair S and T as by the pair P and V. Just as we were able to represent a reversible process as a line joining up a succession of equilibrium states on a P–V diagram, we can do the same on a T–S diagram. However the form of the line is very simple for certain useful processes.

Equation [5.4] shows that a reversible adiabatic process is an isentropic one and such a process is represented on a T–S diagram as a straight line parallel to the T axis, as shown in Fig. 5.10. A reversible isothermal process is represented by a straight line parallel to the S axis. Thus, the cycle for a Carnot cycle is a rectangle on a T–S plot. Compare this with Fig. 4.1.

As for any reversible process:

$$Q = \int_R T \, dS$$

the net heat absorbed in a Carnot cycle is given by the area shaded in the figure.

T–S plots are of enormous value in engineering.

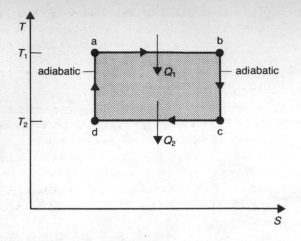

Fig. 5.10 The T–S diagram for a Carnot cycle.

5.8 The central equation of thermodynamics

By combining the first and second laws of thermodynamics, we can obtain the most important equation in thermodynamics.

The differential form of the first law is

$$dU = đQ + đW$$

which is true for both reversible and irreversible processes. For an infinitesimal reversible process, we have

$$đW = -P\,dV \quad \text{and} \quad đQ_R = T\,dS$$

Thus $dU = T\,dS - P\,dV$

or $\quad T\,dS = dU + P\,dV$

We now argue that this equation is true for all processes, whether reversible or not, and not just for reversible processes, as our argument seems to suggest. For, as all quantities in this equation are state functions whose values are fixed by the end points (P, T) and $(P + dP, T + dT)$ of the infinitesimal process, the increments dU, dS and dV are fixed and do not depend on the path joining the end points. Thus any relation between them is independent of whether or not the process is reversible.

This is a significant advance because we now have a general relation between P, V, T and S which holds for all paths between a pair of infinitesimally close equilibrium states, whether or not they are reversible. We call the relation

$$T\mathrm{d}S = \mathrm{d}U + P\mathrm{d}V \qquad [5.10]$$

the *central equation of thermodynamics* or, in view of what we have just said about its generality, the *thermodynamic identity*. The whole of the science of thermodynamics is consequent on this equation, just as the whole of mechanics is consequent on Newton's laws. Because it is an identity, and we do not have to enquire whether the process we are considering is reversible or irreversible, then it follows that the equations derived from it are generally true.

Two remarks should be made at this point, both of which lead to modifications of equation [5.10]. The first is that we have considered only $P\mathrm{d}V$ or volume work; if for example there were also magnetic work, this equation would have to be modified to

$$T\mathrm{d}S = \mathrm{d}U + P\mathrm{d}V - B_0\mathrm{d}\mathcal{M}$$

The second remark is that here we are considering only closed systems where the mass of the system is constant. If this restriction is removed, then additional terms involving the so-called *chemical potential* have to be introduced. Open systems are considered in Chapter 10.

5.9 The entropy of an ideal gas

Although the demonstration of the power of equation [5.10] has to wait until the next chapter, we can give an example here of its use to determine an expression for the entropy of an ideal gas in terms of the volume and temperature.

For an ideal gas, where $U = U(T)$,

$$C_V = \left(\frac{\partial U}{\partial T}\right)_V = \frac{\mathrm{d}U}{\mathrm{d}T}$$

and so equation [5.10] becomes

$$T \, dS = C_V dT + P \, dV$$

Let us consider one mole and use lower-case letters to refer to molar quantities. Using $Pv = RT$, this last equation becomes

$$T ds = c_v dT + \frac{RT}{v} \, dv$$

or $\quad ds = c_v \frac{dT}{T} + R \frac{dv}{v}$

Integrating, we have for the entropy per mole

$$\boxed{s = c_v \ln T + R \ln v + s_0} \qquad [5.11]$$

where s_0 is the integration constant which disappears when entropy differences are taken.

5.10 Entropy, probability and disorder

In our discussion so far, entropy has appeared as a rather mysterious quantity related to the heat flow in reversible processes. Although we have learnt how to use it, we have not really given entropy a physical interpretation. To do this, we have to resort to the microscopic picture, just as we did in Chapter 3 when we interpreted internal energy as the random kinetic and potential energies of the constituent molecules.

How can we describe a system microscopically? We could do so exactly by specifying the position coordinates x, y and z and the momentum components p_x, p_y and p_z for each of the N constituent particles in the system. A given set of these quantities for a particle is represented as a point in the six-dimensional space spanned by x, y, z, p_x, p_y and p_z; this space is called *phase space*. However, we cannot measure x with absolute accuracy but only to within a range Δx; similarly we can measure p_x only to within a range Δp. So we have to be satisfied with less than an exact description by dividing the phase space up into cells of volume $(\Delta x)^3 (\Delta p)^3$ and knowing how many particles are in each cell. We have assumed here that the uncertainty in y and z is also of the order of Δx and that the uncertainty in each of the momentum components is Δp. Now each bulk state is achieved with a different arrangement of the particles amongst the cells and, in general, there will be many different arrangements that can give rise to one particular state. We call the number of arrangements that give rise to a state the *thermodynamic probability*, Ω, of that state. The states with the largest Ω's will be the ones most likely to occur.

This is analogous to the game of craps where two dice are thrown. The most likely score of 7 can be realised in six ways while the least likely scores of 2 or 12 can be realised in only one way (you should work this out).

For our microscopic system, where the number of particles is huge, being of the order of 10^{23}, the thermodynamic probability becomes overwhelmingly large for a particular state, and this will be the observed equilibrium state. An isolated system will move from a state of low thermodynamic probability to the final equilibrium state of maximum thermodynamic probability, consistent with the internal energy U remaining constant. We conclude that

$\Omega \rightarrow$ a maximum

This is our clue as to the meaning of entropy. We remember that, for an isolated system,

$S \rightarrow$ a maximum

while U remains constant. Additionally S is an extensive quantity, so that the entropy of two separate systems is $S_1 + S_2$. If the number of ways of realising the first system is Ω_1 and this is Ω_2 for the second, then the number of ways of realising both systems together is

$\Omega = \Omega_1 \Omega_2$

We see that

$$\boxed{S = k_B \ln \Omega}$$ [5.12]

satisfies these properties. This famous equation is known as the Boltzmann relation and is carved on Boltzmann's tombstone in Vienna. k_B is the so-called Boltzmann constant.

The microscopic viewpoint thus interprets the increase of entropy for an isolated system as a consequence of the natural tendency of the system to move from a less probable to a more probable state.

It is usual to identify Ω as a measure of 'disorder' in the system. This implies that we expect the disorder of an isolated system to increase. To see what this means, let us consider all the particles being in one cell in the phase space. This is a highly ordered arrangement in phase space which can be achieved in only one way with $\Omega = 1$ and $S = 0$. It is a highly ordered arrangement in real space, too, with all the particles being in the same place and moving with identical velocities. The particles will spread out from this highly ordered state, occupying more cells in phase space and lessening the order or increasing the disorder in that space. The thermodynamic probability will increase from 1 to a large value, with the entropy increasing accordingly. It is in this sense that Ω is a measure of disorder.

Let us now be a little more definite and see how there is complete agreement between the macroscopic and microscopic viewpoints in two specific examples.

The entropy change in a free expansion — microscopic approach

We know that there is no temperature change in a free expansion of an ideal gas and so the mean kinetic energy and root mean square momentum $\tilde{p}$ of the molecules remain unaltered. Because it is impossible to draw a six-dimensional phase space, it is difficult at first sight to see what effect this expansion has on the number of phase space cells. However, we can immediately see what happens if we confine ourselves first to a one-dimensional free expansion where the length x is doubled.

Fig. 5.11 shows the phase space for this one-dimensional problem. The length of the momentum side of the phase box must be of the order of $\tilde{p}$ and this is unchanged in the expansion. Doubling the length then doubles the number of cells of size $\Delta x \Delta p$. Extending this argument to three dimensions, the number of cells of cells of size $(\Delta x)^3 (\Delta p)^3$ in the corresponding phase space doubles too if the volume doubles.

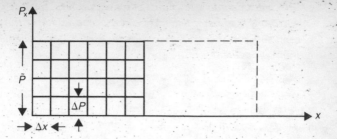

Fig. 5.11 Phase space for a free expansion in a one-dimensional system.

If the number of possible arrangements for fitting the molecules in the cells before expansion is Ω, after expansion this is $2^N \Omega$ because each molecule now has a choice of twice as many cells. Thus

$$\Delta S = k_B \ln(2^N \Omega) - k_B \ln \Omega = k_B \ln 2^N$$

or $\quad \Delta S = Nk_B \ln 2 = nR \ln 2$

for n moles as $N = nN_A$ and $N_A k_B = R$ where N_A is the Avogadro number.

This is exactly the result we obtained earlier as equation [5.6].

The entropy of an ideal gas — microscropic approach

We have previously derived an expression (equation [5.11]) for the entropy of an ideal gas using the central equation of thermodynamics. Let us see if we can derive the same result from microscopic considerations using a simplified argument. We shall consider a monatomic gas where the atoms have translational degrees of freedom only.

The atoms of the gas have to be fitted into the cells of the phase space subject to the following two restrictions:
1. All the atoms have to be contained in a box of volume V.
2. The total energy of the atoms of mass m is fixed at U and is all kinetic, with

$$U = \sum_i \frac{p_i^2}{2m}$$

where the summation is over the N atoms.

The total number of ways Ω of filling up the cells in the phase space is the product of the number of ways Ω_{space} the space cells* of volume Δx^3 can be filled times the number of ways $\Omega_{momentum}$ the different momenta cells of volume Δp^3 can be filled. Thus

$$\Omega = \Omega_{space}\ \Omega_{momentum}$$

Let us calculate Ω_{space}.

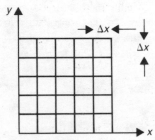

Fig. 5.12 Space cells of volume Δx^3.

In Fig. 5.12 we have drawn the box showing just two dimensions and the space cells. Each atom has $V/\Delta x^3$ distinct locations in the box. Thus

$$\Omega_{space} = \left[\frac{V}{\Delta x^3} \right]^N$$

Let us now calculate $\Omega_{momentum}$. Although each atom is not confined to a finite 'momentum box', they have a root mean square momentum $\widetilde{p}$, given by

$$\frac{\widetilde{p}^2}{2m} = \frac{U}{N}$$

and, for the purpose of this calculation, we may take them as being confined within a momentum box of side $\widetilde{p}$ as shown in Fig. 5.13.

---

* Strictly the use of the word cell in statistical mechanics should be confined to an elementary volume $\Delta x^3\ \Delta p^3$ of phase space.

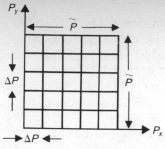

Fig. 5.13 Momentum cells of volume Δp^3.

The number of cells in this momentum box is $(\widetilde{p}/\Delta p)^3$ for each atom. Thus

$$\Omega_{\text{momentum}} \approx \left(\frac{\widetilde{p}}{\Delta p} \right)^{3N}$$

Multiplying these two results,

$$\Omega = \Omega_{\text{space}} \cdot \Omega_{\text{momentum}} \approx \left[\frac{\widetilde{p}^3 V}{\Delta x^3 \Delta p^3} \right]^N$$

However, we have over-counted our ways of filling up the cells because we have assumed that the atoms are *distinguishable*, just as if they are labelled with a number. The two situations depicted in Fig. 5.14 are clearly physically the same.

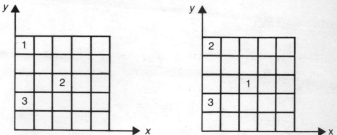

Fig. 5.14 An illustration of two equivalent arrangements in phase space.

There are $N!$ ways of arranging the N identical atoms in a given set of boxes. Thus

$$\Omega_{\text{indistinguishable}} \approx \frac{1}{N!} \left[\frac{\widetilde{p}^3 V}{\Delta x^3 \Delta p^3} \right]^N$$

If N is large, we may use Stirling's approximation for factorials:

$$N! \approx \left(\frac{N}{e}\right)^N$$

So $\quad \Omega_{\text{indistinguishable}} \approx \left[\dfrac{eV\widetilde{p}^3}{N(\Delta x \Delta p)^3}\right]^N$

$$\approx \left[\dfrac{eV(2mU)^{\frac{3}{2}}}{N^{\frac{5}{2}}(\Delta x \Delta p)^3}\right]^N \quad \text{using } \widetilde{p} = \left(\frac{2mU}{N}\right)^{\frac{1}{2}}$$

If we replace the product $\Delta x \Delta p$ with h, this becomes

$$\Omega_{\text{indistinguishable}} \approx \left[\dfrac{eV(2mU)^{\frac{3}{2}}}{N^{\frac{5}{2}}h^3}\right]^N$$

$$\approx \left[\dfrac{eV(2mU)^{\frac{3}{2}}}{(nN_A)^{\frac{5}{2}}h^3}\right]^{nN_A}$$

if we have n moles with $N = nN_A$.

This result is very close to the famous result derived by Sackur and Tetrode using somewhat more rigorous and complicated arguments than the simple ones that we have employed, namely

$$\Omega_{\text{indistinguishable}} = \left[\dfrac{e^{\frac{5}{2}}V(3\pi mU)^{\frac{3}{2}}}{N^{\frac{5}{2}}h^3}\right]^N$$

It is now a simple matter to obtain the entropy:

$$S = k_B \ln \Omega$$

$$= nk_B N_A \left[\ln \frac{V}{n} + \frac{3}{2}\ln \frac{U}{n} + \text{other constant terms}\right]$$

So for one mole,

$$s = R\left(\ln v + \frac{3}{2}\ln u + \text{constant terms}\right)$$

But we know from kinetic theory that

$$u = 3/2\, N_A k_B T$$

for a monatomic gas, so

$$s = R \ln v + 3/2 R \ln T + s_0$$

where s_0 is a constant. As $c_v = 3/2 R$ for a monatomic gas, this is the identical result to equation [5.11] which was obtained using macroscopic ideas.

These two results are a confirmation that entropy is indeed given by

$$S = k_B \ln \Omega \qquad \qquad [5.12]$$

Chapter 6
The thermodynamic potentials and the Maxwell relations

6.1 Thermodynamic potentials

In our formulation of thermodynamics so far, we have seen how the first law allows us to define the internal energy U which is the sum of the random kinetic and potential energies of the component particles of the system. The second law allowed a definition of the entropy S, and we were able to combine both laws into the central equation of thermodynamics

$$T \mathrm{d}S = \mathrm{d}U + P \, \mathrm{d}V \qquad [6.1]$$

which is identically true in that it holds for both reversible and irreversible infinitesimal processes. We shall see the power of this equation in this and the following chapters. However, although its physical interpretation is clear, U is not well suited for the analysis of certain thermodynamic processes and it is convenient to introduce three additional state functions, closely related to U, all of which have the dimensions of energy. These functions are: the enthalpy, H, which we have already met; the Helmholtz free energy, F; and the Gibbs function, G. They do provide a rather more direct link with experiment than can be obtained with the use of U alone. There is a fifth function, the chemical potential, μ, which is useful in discussing the thermodynamics of open systems where the mass of the system is not constant; however, the discussion of μ will be deferred until Chapter 10. The four functions U, H, F and G have a wide applicability throughout thermodynamics and we shall now discuss their properties in turn, some of which we have met before but we shall repeat them here for the sake of completeness. In particular, we shall meet four extremely useful general thermodynamic relations between the four variables P, V, T and S — the four Maxwell relations.

6.2 The internal energy U

Equation [6.1] gives

$$dU = T\,dS - P\,dV$$ [6.2]

As this equation is independent of the type of process used, any relations obtained from it are general ones.

The form of equation [6.2] suggests that we write U in terms of the independent pair of variables S and V as $U = U(S, V)$ where we have used the notation $U(S, V)$ to denote a function of S and V. Hence

$$dU = \left(\frac{\partial U}{\partial S}\right)_V dS + \left(\frac{\partial U}{\partial V}\right)_S dV$$ [6.3]

Comparing equations [6.2] and [6.3],

$$T = \left(\frac{\partial U}{\partial S}\right)_V \quad \text{and} \quad P = -\left(\frac{\partial U}{\partial V}\right)_S$$ [6.4]

This means that, if we know how U depends on its so-called *natural variables* V and S, then we can find T and P. Further, as U is a state function, dU is an exact differential. Using the condition for a differential to be exact (equation [A2.15] of Appendix 2) in equation [6.2],

$$\boxed{\left(\frac{\partial T}{\partial V}\right)_S = -\left(\frac{\partial P}{\partial S}\right)_V}$$ [6.5]

This is our first Maxwell relation. Notice that the natural variables of U, S and V, are the quantities appearing outside the partial differentials.

We can now derive two useful expressions for the heat capacity at constant volume, $C_V = đQ_V/dT$ where the heat has to be put in reversibly. It follows from equation [6.2] that, in a constant volume (or isochoric) process,

$$T\,dS = dU \qquad \text{(isochoric)} \quad [6.6]$$

We also know that, for a reversible process, $T \, dS = đQ$ and so it follows from this and equation [6.6] that, for a reversible isochoric process,

$$đQ_V = dU \qquad \text{(reversible and isochoric)} \qquad [6.7]$$

Hence

$$\boxed{C_V = \left(\frac{\partial U}{\partial T}\right)_V} \qquad \text{[3.6] and [6.8]}$$

and

$$\boxed{C_V = T\left(\frac{\partial S}{\partial T}\right)_V} \qquad [6.9]$$

6.3 The enthalpy H

We have defined H in Chapter 3 as

$$\boxed{H = U + PV} \qquad [3.7]$$

This is a state function as all the quantities on the right-hand side take unique values for each state. Differentiating,

$$dH = dU + P \, dV + V \, dP \qquad [6.10]$$

Using equation [6.1],

$$dH = T \, dS + V \, dP \qquad [6.11]$$

This equation holds for both reversible and irreversible processes by the same argument that we used to show that the central equation of thermodynamics was independent of the type of process, namely that the equation involves only state functions on each side. Again, as this equation is independent of the type of process, any relations obtained from it are general ones.

If we write $H = H(S, P)$,

$$dH = \left(\frac{\partial H}{\partial S}\right)_P dS + \left(\frac{\partial H}{\partial P}\right)_S dP \qquad [6.12]$$

Comparing this equation with equation [6.11],

$$T = \left(\frac{\partial H}{\partial S}\right)_P \quad \text{and} \quad V = \left(\frac{\partial H}{\partial P}\right)_S \qquad [6.13]$$

This means that, if we know H in terms of its natural variables S and P, we can find both the temperature and the volume. Further, using the condition for dH in equation [6.11] to be an exact differential,

$$\boxed{\left(\frac{\partial T}{\partial P}\right)_S = \left(\frac{\partial V}{\partial S}\right)_P} \qquad [6.14]$$

This is our second Maxwell relation, with the natural variables S and P appearing outside the differentials.

The most important property of H is that the change in H is the heat flow in an isobaric reversible process. We obtained this result earlier in Chapter 3 in the infinitesimal form as equation [3.9]:

$$đQ_P = dH \qquad \text{(isobaric and reversible)} \quad [3.9] \text{ and } [6.15]$$

This result can also be obtained directly from equation [6.11] which gives $dH = T\,dS$ for an isobaric process. As $T\,dS = đQ$ for a reversible process, equation [6.15] then follows.

There are two useful expressions for the heat capacity at constant pressure, $C_P = đQ_P/dT$ where the heat has to be added reversibly. From equation [6.15],

$$\boxed{C_P = \left(\frac{\partial H}{\partial T}\right)_P} \qquad [3.10] \text{ and } [6.16]$$

Also, as $T\,dS = đQ$ for a reversible process,

$$\boxed{C_P = T\left(\frac{\partial S}{\partial T}\right)_P} \qquad [6.17]$$

We mentioned in Chapter 3 that the enthalpy change in an isobaric chemical reaction is equal to the heat of reaction. To see this, consider a chemical reaction proceeding so slowly that the pressure above the reacting chemicals is always equal to the pressure of the surroundings, usually the atmospheric pressure P_0. This is shown in Fig. 2.5 where we imagine the reaction taking place in a cylinder with any gas produced gradually pushing out a frictionless and weightless piston. The whole process is irreversible because of the chemical reaction and is just the situation we examined in section 2.5. We saw there that, as the piston is always infinitesimally close to mechanical equilibrium, the work done *by* the expanding gas is, using equation [2.9],

$$W = \int_0^{\Delta V} P \, dV = \int_0^{\Delta V} P_0 \, dV = P_0 \Delta V$$

where ΔV is the volume of gas produced. The first law gives

$$\Delta U = Q - P_0 \Delta V \qquad [6.18]$$

where ΔU is the internal energy change of the whole system and Q is the heat generated in the reaction, or more commonly the *heat of reaction*. As

$$H = U + PV$$

the change in enthalpy ΔH is given by

$$\Delta H = \Delta U + P_0 \Delta V \qquad [6.19]$$

because $P = P_0$ which is constant. From equations [6.18] and [6.19],

$$\Delta H = Q \qquad [6.20]$$

This is the required result. Notice that we have been dealing here with an *irreversible* process and so we could not invoke equation [6.15] which relates to an infinitesimal *reversible* process. The only requirement we made was that the reaction should proceed slowly so that the pressure was always P_0.

6.4 The Helmholtz function F

This state function is designed for problems in which temperature and volume are the important variables and is of enormous value in statistical mechanics. It is defined as

$$F = U - TS \qquad [6.21]$$

For an infinitesimal change,

$$dF = dU - T\,dS - S\,dT \qquad [6.22]$$

Using equation [6.1],

$$dF = -P\,dV - S\,dT \qquad [6.23]$$

Again, this equation is independent of the type of process, so any relations obtained from it are general ones. The natural variables of F are V and T. The form of equation [6.23] suggests that we write

$$F = F(V, T)$$

Hence

$$dF = \left(\frac{\partial F}{\partial V}\right)_T dV + \left(\frac{\partial F}{\partial T}\right)_V dT \qquad [6.24]$$

Comparing coefficients in equations [6.23] and [6.24],

$$P = -\left(\frac{\partial F}{\partial V}\right)_T \quad \text{and} \quad S = -\left(\frac{\partial F}{\partial T}\right)_V \qquad [6.25]$$

Hence, if we know F as a function of volume and temperature, we can find both the entropy and the pressure.

As F is a function of state, dF is an exact differential and the condition for an exact differential in equation [6.23] gives

$$\boxed{\left(\frac{\partial P}{\partial T}\right)_V = \left(\frac{\partial S}{\partial V}\right)_T} \qquad [6.26]$$

This is our third Maxwell relation, with the natural variables V and T appearing outside the differentials.

There are three additional properties of F, each of which we shall discuss in turn.

The maximum work obtainable from a system in which there is no change in temperature

In a purely mechanical system, we are familiar with the idea that the work performed by the system is equal to the decrease in potential energy. In thermodynamics, however, the situation is complicated by the fact that energy can also be exchanged between the system and the surroundings in the form of heat, and we have to look deeper into the relation between the work performed and the change of energy. We shall now derive a very simple expression for the work performed by a system which is in thermal contact with the surroundings at T_0 so that the end points of the process are at T_0, although the intermediate states traversed by the system are *not necessarily* at T_0.

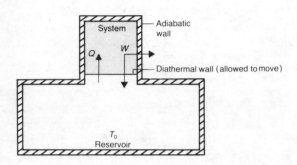

Fig. 6.1 A system in thermal contact with a reservoir at T_0. The system performs work W equal to the decrease in F.

Suppose the system is in thermal contact with a heat reservoir at T_0, as in Fig. 6.1, and let heat Q pass from the reservoir into the system. As we are considering this to be the only flow of heat occurring, we shall surround the system and the reservoir with an adiabatic wall to exclude any other heat flow. The system may *perform* work W; this may be volume work, as its walls are not necessarily rigid, or work in another form such as electrical work. Our principle of increasing entropy gives

$$\Delta S + \Delta S_0 \geqslant 0 \qquad\qquad [6.27]$$

where ΔS and ΔS_0 are the entropy changes of the system and the reservoir. But, as the temperature of the reservoir is unchanged at T_0, $\Delta S_0 = -Q/T_0$ so equation [6.27] becomes

$$\Delta S - Q/T_0 \geqslant 0$$

or $Q - T_0 \Delta S \leqslant 0$ [6.28]

(We have in fact met this equation before, in differential form, as equation [5.7].) Applying the first law to the system,

$$\Delta U = Q - W \qquad [6.29]$$

Substituting for Q, given by equation [6.29], in equation [6.28],

$$\Delta U + W - T_0 \Delta S \leqslant 0$$

or $\Delta(U - TS) + W \leqslant 0$ [6.30]

where we have remembered that the temperature T of the system at the end points is T_0. The expression in brackets in equation [6.30] is just F for the system and so

$$W \leqslant - \Delta F \qquad \text{(no change in } T) \quad [6.31]$$

The equality sign holds for a reversible process which we know produces the maximum amount of work. Hence, *in a process in which the end point temperatures are the same as the surroundings, the maximum work obtainable is equal to the decrease in the Helmholtz function.* Such processes are commonly encountered. Because of this relation, F is often called the *Helmholtz free energy.* Indeed, it is often given the symbol A after the German for work, *Arbeit.*

It should be emphasised that T does *not* have to be held constant in an isothermal process at T_0 for equation [6.31] to hold but has to assume this value *only* at the end points. A less general proof of equation [6.31] follows immediately from equation [6.23] for the special case of an isothermal process, as then $- dF = P \, dV = dW_{rev}$. Unfortunately, this is the proof usually given elsewhere in other texts.

The equilibrium condition for a system in thermal contact with a heat reservoir and held at constant volume

There is a simple argument to show that the equilibrium condition for a system in thermal contact with a heat reservoir and held at constant volume is one of minimum F. To see this, let us consider the system, shown in Fig. 6.2, which is in thermal contact, via a diathermal wall, with a reservoir at a temperature T_0 so that its temperature T is also T_0. Let the volume of the system be fixed at V. The combined system of the system and the reservoir is surrounded by an adiabatic wall.

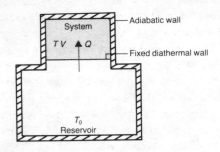

Fig. 6.2 A system in thermal contact with a reservoir at T_0. The volume of the system is fixed. The condition for equilibrium is that F is a minimum.

If the system is a *simple* one consisting of a fixed mass of homogeneous material, then specifying the temperature and volume of the system *fixes its state* and there is complete or thermodynamic equilibrium; clearly, no further changes can occur. However, other internal degrees of freedom may exist within a more *complex* system so that it may not be in a state of thermodynamic equilibrium, even though we have specified the temperature and the volume. An irreversible process can then occur involving the transfer of heat Q from the reservoir to the system as it tends towards equilibrium.

As an example of what is meant by an extra degree of freedom, we can imagine a chemical reaction occurring within the system

$$A + B \rightarrow AB$$

with the ratio of the amount of product AB to the amount of reactant A or B being a variable quantity which changes as equilibrium is approached. Alternatively there could be a mixture of ice and water within the system, with the ratio of ice to water varying until equilibrium is reached. In both of these examples, heat is transferred between the system and the reservoir; this is the heat of reaction for the chemical reaction and the latent heat for the melting of the ice.

We shall now argue that, in our irreversible process taking the system and the reservoir towards equilibiurm, ΔF for the system is necessarily negative (and ultimately zero). However, we must first examine exactly what we mean by ΔF. To help us with our discussion, we shall consider the specific example of a mixture of ice and water reaching equilibrium. The concept of a well defined final F causes us no difficulty because the final state is an equilibrium state and we simply add the separate F's for the ice and water. How, though, can we talk of a free energy for the initial non-equilibrium state? The argument is exactly the same as the one we employed on page 84 for the entropy of a bar with a temperature gradient and in a non-equilibrium state. At the beginning of our process, and also at any stage, we may imagine separating the ice and water so that we again have an equilibrium situation. We may then add up their separate F's as before to give a definite F for the system. In this way, we may talk of a ΔF for the process. Why, then, is $\Delta F \leqslant 0$?

To see this, let us *first* assume that the process takes place isothermally at T_0 and at the constant volume V. As the combined system of the system and the thermal reservoir is thermally isolated from the rest of the universe by the adiabatic wall, it follows from the principle of increasing entropy, exactly as on page 107 of this section that

$$Q - T_0 \Delta S \leqslant 0 \qquad [6.28]$$

where ΔS is the entropy change of the system. Applying the first law to the system and remembering that no work is done as we are keeping the volume constant at V,

$$\Delta U = Q \qquad [6.32]$$

(We are excluding from our discussion any other types of work such as magnetic work.) Substituting for Q in equation [6.28],

$$\Delta U - T_0 \Delta S \leqslant 0$$

$$\text{or} \quad \Delta U - \Delta(TS) \leqslant 0 \qquad [6.33]$$

as the temperature T is fixed at the constant value of T_0 for the process under consideration. Rewriting equation [6.32],

$$\Delta(U - TS) \leqslant 0$$

or $\quad \Delta F \leqslant 0$ [6.34]

But we may now argue in the usual way that, as F is a state function, the decrease ΔF in the Helmholtz function is the same for any process between a *given pair* of equilibrium states. Thus, equation [6.34] holds in general for any process between a pair of states at the same temperature and volume, and not just for the special isochoric and isothermal process that we have just considered. It is important to realise that the temperature and volume do not have to be fixed at T_0 and V during the process for equation [6.34] to hold, but that they have these values *only* at the end points. Indeed, as we have seen in our example of a chemical reaction, the temperature will almost certainly change during the reaction before settling back to T_0 when it is over.

We conclude from equation [6.34] that, for a system of constant volume in thermal contact with a heat reservoir, spontaneous (not externally induced) changes will occur in the direction of decreasing F. Further, at equilibrium, where any infinitesimal changes induced are reversible and the equality sign has to be taken in equation [6.34],

$$dF = 0$$ [6.35]

and we have a minimum in the Helmholtz function. Hence

The condition for thermodynamic equilibrium in a system in thermal contact with a heat reservoir and maintained at constant volume is that the Helmholtz function is a minimum.

The bridge equation between thermodynamics and statistical mechanics

It is through the Helmholtz free energy that an important link between statistical mechanics and thermodynamics is made. To discuss this we have to meet the concept of the *partition function* z. Although z can be defined for a classical system, where the energy can take a continuous range of values, we shall restrict our discussion to the quantum case where each microscropic particle of the system can be in one of a whole series of quantum states (not to be confused with the words thermodynamic state) with energies, or energy levels

$$\epsilon = \epsilon_1, \epsilon_2, \epsilon_3 \ldots \epsilon_i \ldots$$

as shown in Fig. 6.3. Some quantum states may have the same energy; the number of quantum states with the same energy level is called the *degeneracy*, g, of that energy level.

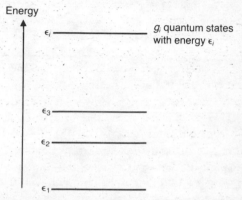

Fig. 6.3 A series of energy levels ϵ_i for the particles comprising a system. The degeneracy of the ith level is g_i. The single-particle partition function is $z = \sum_i g_i e^{-\epsilon_i/k_B T}$.

The single particle partition function is defined as:

$$z = \sum_i g_i \, e^{-\epsilon i/k_B T}$$

It is a standard result of statistical mechanics that

$$F = -N k_B T \ln z \qquad [6.36]$$

for the N particles making up the system, providing that they are only weakly interacting and are distinguishable from each other. This very important relation is known as the bridge equation for the so-called canonical distribution in statistical mechanics. Once the partition function has been evaluated, F can be obtained using equation [6.36] and, from this, S and P can be evaluated using equation [6.25]. The thermodynamic description of the system is then known.

As a simple illustration of these ideas, let us calculate the entropy of a two-level system of N weakly interacting distinguishable particles, where each particle can exist in one of two quantum states with energies 0 and ϵ. Such a system could be a collection of electron spins, as in a paramagnetic salt, in a magnetic field. Then,

$$z = e^{-0/k_BT} + e^{-\epsilon/k_BT} = 1 + e^{-\epsilon/k_BT}$$

so
$$F = -Nk_BT \ln(1 + e^{-\epsilon/k_BT})$$

Thus
$$S = -\left(\frac{\partial F}{\partial T}\right)_V = Nk_B \left[\ln(1 + e^{-\epsilon/k_BT}) + \frac{\epsilon}{k_BT} \cdot \frac{1}{(1 + e^{\epsilon/k_BT})} \right]$$

which gives the variation of entropy with temperature. At low temperatures ($k_BT \ll \epsilon$) this expression approximates to $S = 0$ while at high temperatures ($k_BT \gg \epsilon$) it yields $S = Nk_B \ln 2$. These results are consistent with equation [5.12] because, in the high temperature limit, each particle has sufficient thermal energy to be able to occupy either of the two energy levels and so $\Omega = 2^N$ giving our $S = Nk_B \ln 2$ result again. Also, in the low temperature limit, each particle can occupy only the lowest level so $\Omega = 1^N = 1$ with $S = 0$. Our partition function approach has told us much more than this: it has given us the actual temperature variation of S.

In general, the partition function is an invaluable tool in the calculation of the bulk thermodynamic properties of a system using statistical mechanics, especially when more complicated systems are considered than the one here.

6.5 The Gibbs function G

This state function is designed for use in problems where pressure and temperature are the important variables. It is of enormous importance in chemistry and in the study of systems where there is a mixture of two phases of a substance, such as a mixture of ice and water; these systems will be discussed in Chapter 9. The Gibbs function is defined as

$$G = H - TS$$ [6.37]

For an infinitesimal change,

$$dG = dH - T\,dS - S\,dT$$

or $$dG = dU + P\,dV + V\,dP - T\,dS - S\,dT$$

using equation [3.7]. It follows from equation [6.1] that:

$$dG = V\,dP - S\,dT$$ [6.38]

Again this equation is independent of whether or not the process is reversible and so any relations derived from it are general ones. The natural variables of G are P and T. Equation [6.38] shows that, in any process that takes place at constant T and P, the Gibbs function is unchanged.

The form of equation [6.38] suggests that we write

$$G = G\,(P, T)$$

Hence

$$dG = \left(\frac{\partial G}{\partial P}\right)_T dP + \left(\frac{\partial G}{\partial T}\right)_P dT$$ [6.39]

Comparing the coefficients in equations [6.38] and [6.39],

$$V = \left(\frac{\partial G}{\partial P}\right)_T \quad \text{and} \quad S = -\left(\frac{\partial G}{\partial T}\right)_P$$ [6.40]

Hence, if we know G as a function of P and T, we can find the volume and the entropy.

As G is a state function, dG is an exact differential. Using the condition for an exact differential in equation [6.38],

$$\left(\frac{\partial V}{\partial T}\right)_P = -\left(\frac{\partial S}{\partial P}\right)_T$$ [6.41]

This is our fourth Maxwell relation with the natural variables P and T appearing outside the partial differentials.

There are two further important properties of G which we shall now discuss.

The equilibrium condition for a system in thermal and mechanical contact with a heat and pressure reservoir

Just as we saw that a system held at constant volume and in thermal contact with a heat reservoir assumes an equilibrium state which is one of minimum F, we shall show again, using the principle of increasing entropy, that a system in thermal and mechanical contact with a heat and pressure reservoir moves to an equilibrium state of minimum G. By such a reservoir we mean one which is so large that its temperature and pressure remain unchanged, whatever is done to it. This result for the minimisation of G is of enormous importance because the conditions we have described are exactly the ones encountered in many natural processes where the surrounding atmosphere at P_0 and T_0 acts as a pressure and heat reservoir. As examples of such processes we could include most chemical reactions, and phase changes such as ice melting into water in a beaker open to the atmosphere (we shall return to phase changes in Chapter 9).

Let us consider the system shown in Fig. 6.4 which is in contact with a heat and pressure reservoir at T_0 and P_0 via a

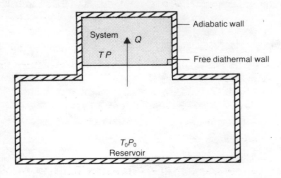

Fig. 6.4 A system in contact with a reservoir at T_0 and P_0. The pressure and temperature of the system at the end points are maintained at P_0 and T_0 because the intervening wall is diathermal and can move freely. The condition for equilibrium is that G is a minimum.

diathermal piston, which is weightless, free and frictionless. As in our discussion of F, let there be some internal degree of freedom in the system so that it is not in thermodynamic equilibrium. As before, let heat Q pass from the reservoir to the system in a spontaneous process from an initial non-equilibrium state at P_0 and T_0 to a final equilibrium state also at P_0 and T_0. We can define a change ΔG in the Gibbs function in exactly the same way that we defined ΔF in section 6.3. We can now argue that ΔG must be negative if the process is irreversible.

Let us *for a moment* assume that this process is both isothermal and isobaric. Because the combined system of the system and the reservoir is thermally isolated from the rest of the universe, it follows from the principle of increasing entropy, exactly as in section 6.4, that

$$Q - T_0 \Delta S \leqslant 0 \qquad [6.28]$$

where ΔS is the entropy change of the system. Applying the first law to the system and remembering that the work this does is $P_0 \Delta V$ where ΔV is its volume change,

$$\Delta U = Q - P_0 \Delta V \qquad [6.42]$$

(Again, we have excluded other types of work, such as magnetic work.) Substituting for Q in equation [6.28],

$$\Delta U + P_0 \Delta V - T_0 \Delta S \leqslant 0$$

or $\quad \Delta U + \Delta(PV) - \Delta(TS) \leqslant 0 \qquad [6.43]$

as the pressure and temperature of the system are fixed at the constant values of P_0 and T_0 for the process under consideration. Rewriting equation [6.43],

$$\Delta(U + PV - TS) \leqslant 0 \qquad [6.44]$$

or $\quad \Delta G \leqslant 0 \qquad [6.45]$

But we now argue that, as G is a state function, the decrease in G is the same for *any* process between this pair of end point states as it is for the *particular* isothermal and isobaric process we have just considered. Thus equation [6.45] holds in general for a process between a pair of states at the same temperature and pressure as the surroundings. It is very important to realise that the temperature and pressure do not have to remain fixed at T_0 and P_0 during the process of equation [6.45] to hold, only that they assume these values at the end points. Indeed, in a chemical reaction the temperature, and possibly the pressure, will almost certainly change before settling back to the ambient temperature and pressure at the final equilibrium state.

We conclude from equation [6.45] that, for a system in contact with a heat and pressure reservoir, spontaneous changes will occur in the direction of decreasing G. At equilibrium, where any infinitesimal changes induced are reversible, the equality sign has to be taken in equation [6.45] and

$$\mathrm{d}G = 0 \qquad\qquad [6.46]$$

so we have a minimum in the Gibbs function. Hence:

The condition for thermodynamic equilibrium in a system in thermal and mechanical contact with a heat and pressure reservoir is that the Gibbs function is a minimum.

Let us examine here the significance of this result for a chemical reaction. In a typical reaction,

A + B → C + D

the reactants A and B are initially at room or ambient temperature; they react, perhaps giving out heat, with the temperature rising; and then the products of the reaction C and D cool back to the room temperature again. The pressure is kept at the pressure of the surroundings, usually atmospheric pressure. These are just the conditions we have been considering. The reaction will proceed if ΔG for the process is negative. Let us be a little more specific by considering two important biological reactions.

The oxidation of glucose is represented by the following reaction:

$$C_6H_{12}O_6 + 6O_2 \rightarrow 6CO_2 + 6H_2O$$

The values for the Gibbs function for all the chemicals are tabulated. At 25 °C the change in G for the reaction is -2.9×10^6

J mol^{-1} so the reaction will proceed in the direction indicated. However, thermodynamics does *not* tell us the *rate* at which the reaction proceeds, only its *direction*. In practice, an enzyme catalyst is required to speed up this reaction.

Photosynthesis, the formation of carbohydrate, is represented by the following reaction:

$$CO_2 + H_2O \rightarrow CH_2O + O_2$$

ΔG for this reaction is $+ 4.7 \times 10^5$ J mol^{-1}, so the reaction will not proceed spontaneously; the radiant energy of sunlight is required to drive the reaction forward.

It is instructive to conclude this section by looking again, in rather more general terms, at the underlying reasons for the minimisation of G for a closed system in thermal and mechanical contact with the surroundings at T_0 and P_0. We know that the entropy of the system and the surroundings can only increase or stay the same:

$$\Delta S^{\text{universe}} = \Delta S + \Delta S_0 \geqslant 0$$

This is the fundamental idea in our argument. We have seen in equation [6.43] that this implies

$$\boxed{\Delta H - T_0 \Delta S \leqslant 0}$$

because $\Delta U + \Delta(PV) = \Delta H$ and $\Delta(TS) = T_0 \Delta S$. Although the state functions S and H in this inequality refer to the system, it is important to remember that the inequality tells us about the net entropy change in the *system and the surroundings*. Now the interdependent entropy and enthalpy changes ΔH and ΔS of the system which occur in a process may be positive or negative and the necessary requirement for the process to proceed is that the inequality is satisfied so that the entropy of the universe increases. Clearly this is so if ΔS is positive and ΔH is negative; such a process may then proceed of its own accord (i.e. spontaneously). Conversely, if ΔH is positive and ΔS is negative, the process cannot proceed. However, if ΔH and ΔS have the same sign, then whether or not the process proceeds depends on which of the two terms, ΔH and $T_0 \Delta S$ is dominant. There is competition between them and a balance will be struck when $T_0 \Delta S = \Delta H$ with $\Delta S^{\text{universe}} = 0$.

This balance at equilibrium appears as a minimisation of G which contains both H and S. We should contrast these ideas with those from statics where, at equilibrium, it is the simple potential energy which is minimised. In thermodynamics, we have allowed for the additional complication of adding energy to the system in the form of heat rather than just by work, as in statics. The potential function G looks after the competing changes in H and S and *automatically* includes the entropy changes of *both* the system and the surroundings for our process where the end points are at the ambient T_0 and P_0. Were we to have considered the less important case of a system at a fixed volume in thermal contact with a heat reservoir at T_0, we would have found that there is competition between ΔU and $T_0 \Delta S$ to give a minimum in F at equilibrium.

Finally we note that it is the $T_0 \Delta S$ term which is dominant at high temperatures, while the enthalpy term dominates at low temperatures. To see what this means, let us consider a process represented by

$$A \rightleftarrows B$$

where the double arrow implies that the process between two states A and B can go either way under suitable conditions. Suppose that the process A → B has $\Delta S > 0$, which is favourable for the change A → B, but that $\Delta H > 0$, which is unfavourable for a change in that direction. Of course, this latter condition is favourable to the *reverse* process B → A. As an example of such a process, we could consider ice melting into water, with A being ice and B being water. As ice requires latent heat in order for it to melt, ΔH is positive. {This is physically reasonable: $\Delta H = \Delta U + \Delta(PV) \approx \Delta U$ for a solid–liquid transition as the $\Delta(PV)$ term is generally small compared with the ΔU term, because of the small volume change, and ΔU is positive because the potential energy of the molecules increases when they break free from their positions in the solid.} Now ΔS is also positive because we are going from an ordered state to a more disordered one. At low temperatures, the ΔH term wins, so ΔG is positive and the process A → B does not occur. However, at high temperatures, the $T_0 \Delta S$ term wins so that ΔG is now negative and the process A → B goes ahead. This certainly is true for our example with ice at one atmosphere remaining ice at temperatures less than 0 °C and melting into water at higher temperatures. The stable condition of ice coexisting with water occurs when the ΔH term is exactly balanced by the $T_0 \Delta S$ term at 0 °C with $\Delta G = 0$ (we shall return to this point in Chapter 9). These ideas, with a particular emphasis on chemical systems, are beautifully discussed in the book on chemical thermodynamics by Warn (see Appendix 6).

Useful work

We have seen in section 6.4 that, in a process in which the initial and final temperatures are equal to the surroundings and where the heat transferred is between the system and the surroundings only, the maximum work that can be obtained from a system is equal to the decrease in F. Let us consider a gas in a cylinder expanding through a volume change ΔV and performing some external work W. This could be the lifting of a weight, as illustrated in Fig. 6.5(a) or the turning of the shaft of an electrical generator for example. In performing this work, the gas also does *useless work* $P_0 \Delta V$ against the surrounding atmosphere at the pressure P_0 and temperature T_0. Another important example of useless work occurs in an electrolytic cell where, in addition to the electrical work delivered at the rate $I^2 R$ when a current I is passed through a resistance R, the cell also performs useless work because any gases produced in the electrolytic reaction have to push back the atmosphere. This is illustrated in Fig. 6.5(b).

There is an elegant way of subtracting off this useless work from the total work performed by a system which is allowed to expand between a pair of states, both at P_0 and T_0, and where any heat transferred is between the system and the surroundings only. Our cell illustrated in Fig. 6.5(b) is an example of such a process; however, we have to exclude our gas–weight system of Fig. 6.5(a) because the end point pressures are not P_0, even though the end point temperatures are T_0.

Before proceeding further, let us briefly examine the requirement that the end points be at P_0 and T_0. Clearly, in a *simple* system this means that there can be no change in volume and so no volume work can be performed. For a change of volume to

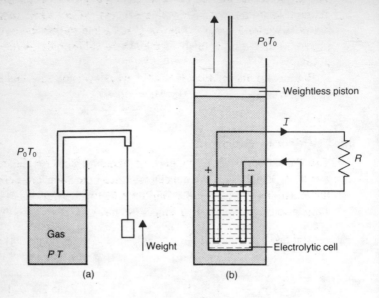

Fig. 6.5 Two illustrations of useful work. In each case the system has to expand against the surroundings and, in doing so, has to perform an amount $P_0 \Delta V$ of useless work.

occur under these conditions, there must be some additional degree of freedom within the system, as would be allowed, for example, by an internal chemical reaction.

For our process, assuming it to be reversible,

$$W = -\Delta F \qquad [6.31]$$

as the end point temperature are the same. Also

$$W^{\text{useful}} = W - P_0 \Delta V = -\Delta F - P_0 \Delta V \qquad [6.47]$$

or $\quad W^{\text{useful}} = -\Delta(F + P_0 V) = -\Delta(F + PV) \qquad [6.48]$

as the end point pressures are both P_0. As can be seen from the defining equations [6.21] and [6.37], $G = F + PV$, and it follows from equation [6.48] that

$$W^{\text{useful}} = -\Delta G \qquad \text{(end points at } P_0 T_0) \qquad [6.49]$$

The decrease in the Gibbs function then gives the maximum amount of useful work for such a process. Notice again that we did not require T and P to be fixed at T_0 and P_0 throughout the process, but required them to take these values only at the end points.

Because of its relation to useful work, G is often called the *Gibbs free energy*.

6.6 The Maxwell relations

This has been a difficult chapter, containing some complex arguments. It is helpful if we collect together here what are perhaps the most important results, our four Maxwell relations, which are relations between P, V, T and S. We have:

$$\left(\frac{\partial T}{\partial V}\right)_S = -\left(\frac{\partial P}{\partial S}\right)_V \qquad \text{(from } U\text{)} \quad [6.5]$$

$$\left(\frac{\partial T}{\partial P}\right)_S = \left(\frac{\partial V}{\partial S}\right)_P \qquad \text{(from } H\text{)} \quad [6.14]$$

$$\left(\frac{\partial P}{\partial T}\right)_V = \left(\frac{\partial S}{\partial V}\right)_T \qquad \text{(from } F\text{)} \quad [6.26]$$

$$\left(\frac{\partial V}{\partial T}\right)_P = -\left(\frac{\partial S}{\partial P}\right)_T \qquad \text{(from } G\text{)} \quad [6.41]$$

We shall constantly be using these four relations in the derivation of our thermodynamic relationships. Although they are easy to derive, and we *should all be able to do so*, it is extremely valuable to be able to recall them. It is an almost impossible task to remember them as each partial differential has six possible combinations of the variables and one's memory will almost certainly fail. Fortunately there is a very simple mnemonic which is invaluable. Let us join the Society for the Prevention of Teaching of Vectors, written as

$$\begin{array}{ccc} & S & \\ \overline{} & & \\ P & & V \\ & T & \end{array}$$

The thermodynamic potentials and the Maxwell relations   121

All one has to do is to go round this array cyclically in opposite directions. For example, the second Maxwell would follow from starting at T as:

There are four possible starting points and so we can produce the four Maxwell equations. The only complication is the minus sign. When S and P appear together in a partial differential, a minus sign must be put outside — hence the minus sign at the top left of the mnemonic.

There are other mnemonics which give much more than the Maxwell relations, but it is the author's experience that one needs a mnemonic to remember them! We shall see that everything can be worked out from just a few basic equations and this mnemonic; there will be no need to remember complicated formulae, and one need never get lost in a complicated thermodynamic derivation involving seemingly endless partial differentials — the route through the maze is always well signposted so that the direction in which to go is clear. Finally, we must emphasise that our mnemonic is only just that. Although it is invaluable, we must not use it as a proof of the Maxwell relations, especially in an examination!

6.7 An example of the use of the Maxwell relations

$$
\boxed{\begin{array}{c} P_i \\ T_i = T \end{array}} \longrightarrow \boxed{\begin{array}{c} P_f \\ T_f = T \end{array}}
$$

Suppose we have a block of metal and we squash it reversibly and isothermally, at the temperature T, from a pressure P_i to a pressure P_f. Heat will flow out of the metal. Let us calculate how much heat flows out of the block.

The problem always in a thermodynamics problem is knowing

where to start, and the answer is usually obvious by going back to the appropriate basic equation. Here, we write down our basic equation for heat flow in a reversible process:

$$\text{đ}Q_R = T\,dS \qquad [5.4]$$

Hence, if we can find dS in terms of quantities we know and then integrate, we have solved the problem. As in the example given in Chapter 1, we express S as a function of the state variables in which we are given the changes, in this case P and T. That is

$$S = S(T,P)$$

Hence

$$dS = \left(\frac{\partial S}{\partial P}\right)_T dP + \left(\frac{\partial S}{\partial T}\right)_P dT$$

so $\quad \text{đ}Q_R = T\,dS = T\left(\frac{\partial S}{\partial P}\right)_T dP + T\left(\frac{\partial S}{\partial T}\right)_P dT$

where the second term is zero as the process is isothermal. The first partial differential containing S can hopefully be changed into a more familiar expression involving P, V and T and quantities we know by using a Maxwell relation. An application of our mnemonic shows that:

$$
\begin{array}{ccc}
 & S & \\
- & & \\
P & & V \\
 & T &
\end{array}
\qquad
\left(\frac{\partial S}{\partial P}\right)_T = -\left(\frac{\partial V}{\partial T}\right)_P
\qquad [6.41]
$$

(which of course is the Maxwell relation obtained from G). Hence

$$\text{đ}Q = -T\left(\frac{\partial V}{\partial T}\right)_P dP = -TV\beta\,dP$$

Integrating,

$$Q = -\int_{P_i}^{P_f} TV\beta\,dP \approx -TV\beta \int_{P_i}^{P_f} dP$$

$$\approx -TV\beta\,(P_f - P_i)$$

The thermodynamic potentials and the Maxwell relations 123 where we have assumed that there is only a small change of volume in the compression and the expansion coefficient β to be constant. The — sign indicates that heat flows out.

Virtually all problems in thermodynamics can be tackled using this approach.

Chapter 7

Some general thermodynamic relations

We are at the stage in our argument where we can apply our ideas to obtain some very powerful general results. These results all follow very simply from the basic equations, but it is important to think the right way so not to get lost in the derivations. This method of thinking and of 'reading the signposts' will be brought out in the following sections of this chapter.

7.1 The difference in the heat capacities, $C_P - C_V$

We saw in section 3.6 that, for an ideal gas,

$$C_P = C_V + nR \qquad [3.12]$$

We shall now generalise this result for any system with the state variables P, V and T.

Our starting point comes from either of the two basic relations for the principal heat capacities:

$$C_V = T \left(\frac{\partial S}{\partial T} \right)_V \qquad [6.9]$$

and $\quad C_P = T \left(\frac{\partial S}{\partial T} \right)_P \qquad [6.17]$

If we choose equation [6.9], its form suggests we write

$$S = S(T, V) \qquad [7.1]$$

so that

$$dS = \left(\frac{\partial S}{\partial T}\right)_V dT + \left(\frac{\partial S}{\partial V}\right)_T dV \qquad [7.2]$$

As we need to obtain C_P, let us 'divide' equation [7.2] all through by dT at constant pressure and multiply by T^*. This gives C_P on the left-hand side and C_V on the right-hand side as

$$T\left(\frac{\partial S}{\partial T}\right)_P = T\left(\frac{\partial S}{\partial T}\right)_V + T\left(\frac{\partial S}{\partial V}\right)_T \left(\frac{\partial V}{\partial T}\right)_P \qquad [7.3]$$

or $\quad C_P = C_V + T\left(\frac{\partial S}{\partial V}\right)_T \left(\frac{\partial V}{\partial T}\right)_P \qquad [7.4]$

Our aim is to produce a relation between C_P and C_V in terms of quantities that we can measure, in particular β and K where

$$\beta = \frac{1}{V}\left(\frac{\partial V}{\partial T}\right)_P \qquad [2.1]$$

and $\quad K = -V\left(\frac{\partial P}{\partial V}\right)_T = \frac{1}{\kappa} \qquad [2.3]$

The second partial differential in the second term on the right of equation [7.4] is essentially β. The first partial differential we do not recognise, so let us see if we can remove the entropy by applying a Maxwell relation. Our mnemonic immediately tells us that we can do so as

$$\left(\frac{\partial S}{\partial V}\right)_T = \left(\frac{\partial P}{\partial T}\right)_V \qquad [6.26]$$

$\begin{array}{ccc} & S & \\ - & & \\ P & & V \\ & T & \end{array}$

Hence equation [7.4] becomes

$$C_P = C_V + T\left(\frac{\partial P}{\partial T}\right)_V \left(\frac{\partial V}{\partial T}\right)_P \qquad [7.5]$$

But the first partial differential in equation [7.5] has P, V and T in the wrong order to use K or β directly, so we have to employ the cyclical rule of Appendix 2:

* Strictly, we should write $dS = (\partial S/\partial T)_P \, dT + (\partial S/\partial P)_T \, dP$ and $dV = (\partial V/\partial T)_P \, dT + (\partial V/\partial P)_T \, dP$ in equation [7.2], then compare coefficients of dT.

$$\left(\frac{\partial P}{\partial T}\right)_V \left(\frac{\partial V}{\partial P}\right)_T \left(\frac{\partial T}{\partial V}\right)_P = -1$$

or $\quad \left(\frac{\partial P}{\partial T}\right)_V = -\left(\frac{\partial P}{\partial V}\right)_T \left(\frac{\partial V}{\partial T}\right)_P$

Hence, equation [7.5] becomes

$$C_P = C_V - T\left(\frac{\partial V}{\partial T}\right)_P^2 \left(\frac{\partial P}{\partial V}\right)_T$$

or $\quad \boxed{C_P = C_V + T\beta^2 KV}$ [7.6]

which is our final result. It should *not* be remembered, as it is so easy to derive.

Equation [7.6] is a surprising result, relating apparently unconnected parameters — heat capacity, expansivity and elastic modulus. It is also a very useful relation because experiment usually measures C_P while theory gives C_V, and so the two can be compared. Also, as K is positive for all known substances, we see that $C_P > C_V$ in general. Finally, it is left to the reader to show that equation [7.6] reduces to equation [3.12] for an ideal gas.

7.2 Evaluation of $\left(\frac{\partial C_V}{\partial V}\right)_T$ and $\left(\frac{\partial C_P}{\partial P}\right)_T$

For an ideal gas, we know that $U = U(T)$ only, and so $C_V = (\partial U/\partial T)_V = dU/dT$ is a function of temperature only too. In fact, for an ideal monatomic gas, $U = 3N_A k_B T/2$ per mole and so C_V is a constant. The ideal gas is typical of many systems, in that theory often tells us how the internal energy, and hence C_V, varies with T. However we are prompted to ask the question of how C_V varies with V in general; in other words what is $(\partial C_V/\partial V)_T$?

We have

$$C_V = T\left(\frac{\partial S}{\partial T}\right)_V$$ [6.9]

so $\quad \left(\dfrac{\partial C_V}{\partial V}\right)_T = T \left(\dfrac{\partial}{\partial V}\right)_T \left(\dfrac{\partial S}{\partial T}\right)_V$

or $\quad \left(\dfrac{\partial C_V}{\partial V}\right)_T = T \left(\dfrac{\partial}{\partial T}\right)_V \left(\dfrac{\partial S}{\partial V}\right)_T$ [7.7]

as the order of differentiation with respect to V and T may be interchanged in the second order partial differential. Applying the Maxwell relation

$$\left(\frac{\partial S}{\partial V}\right)_T = \left(\frac{\partial P}{\partial T}\right)_V \qquad [6.26]$$

to equation [7.7],

$$\left(\frac{\partial C_V}{\partial V}\right)_T = T \left(\frac{\partial}{\partial T}\right)_V \left(\frac{\partial P}{\partial T}\right)_V$$

or $\quad \boxed{\left(\dfrac{\partial C_V}{\partial V}\right)_T = T \left(\dfrac{\partial^2 P}{\partial T^2}\right)_V}$ [7.8]

If one knows the equation of state, equation [7.8] may be integrated to tell one how C_V varies with V. It is left as a question (question 2, Chapter 7, Appendix 4) for the reader to show that C_V for a van der Waals gas is a function of T only.

It is also left as a question (question 1, Chapter 7, Appendix 4) for the reader to derive the analogous result:

$$\boxed{\left(\frac{\partial C_P}{\partial P}\right)_T = - T \left(\frac{\partial^2 V}{\partial T^2}\right)_P} \qquad [7.9]$$

Again, these relations should *not* be remembered as they are so easy to derive; it is sufficient to remember that simple expressions exist for

$$\left(\frac{\partial C_V}{\partial V}\right)_T \quad \text{and} \quad \left(\frac{\partial C_P}{\partial P}\right)_T$$

7.3 The energy equation

We know that $U = U(T)$ only for the special case of an ideal gas. However, in general U will be a function of volume too, so let us see if we can find a relation expressing this dependence. In other words, can we find an expression for $(\partial U/\partial V)_T$ in terms of P, V and T?

We have the basic relation

$$dU = T\,dS - P\,dV \qquad [6.1]$$

so

$$\left(\frac{\partial U}{\partial V}\right)_T = T\left(\frac{\partial S}{\partial V}\right)_T - P$$

or

$$\boxed{\left(\frac{\partial U}{\partial V}\right)_T = T\left(\frac{\partial P}{\partial T}\right)_V - P} \qquad [7.10]$$

using the Maxwell relation

$$\left(\frac{\partial S}{\partial V}\right)_T = \left(\frac{\partial P}{\partial T}\right)_V \qquad [6.26]$$

Equation [7.10] is a very powerful equation and is known as the *energy equation*. If we know the equation of state for the system, we can use equation [7.10] to gain information about the internal energy.

As an example of its application, let us consider an ideal gas with the equation of state

$$PV = nRT$$

so

$$\left(\frac{\partial U}{\partial V}\right)_T = T\frac{nR}{V} - P = 0$$

This tells us that U is not an explicit function of V but only of T:

$$U = U(T) \qquad [7.11]$$

It might be asked why U is not an explicit function of P. It cannot be for the following reason. Suppose

$$U = f(T, P) \qquad [7.12]$$

where f is a function. Then, using the equation of state, we could substitute for P in equation [7.12] to obtain a new function, g, where

$$U = g(T, V) \qquad [7.13]$$

which, as we have just shown from the energy equation, is untrue. We conclude then that equation [7.12] is false, and

$$U = U(T) \qquad [7.11]$$

only for an ideal gas — the result we gave in section 3.6.

There is a second energy equation which is used less frequently than the first (equation [7.10]). It is left for the reader to show, in question 3, Chapter 7, Appendix 4, that

$$\boxed{\left(\frac{\partial U}{\partial P}\right)_T = - \left\{ T\left(\frac{\partial V}{\partial T}\right)_P + P\left(\frac{\partial V}{\partial P}\right)_T \right\}} \qquad [7.14]$$

Again, one should not attempt to remember the energy equations. It is sufficient to remember that they exist and the form of the left-hand sides, which is easy if one thinks along the lines of the introduction to this section.

7.4 The ratio of the heat capacities, C_P/C_V

Just as we saw that there is a useful relation for the difference in the principal heat capacities, so there is a useful expression for their ratio. We shall show that

$$\frac{C_P}{C_V} = \frac{\kappa_T}{\kappa_S}$$

where κ_T is the usual isothermal compressibility and κ_S is the adiabatic compressibility:

$$\kappa_T = -\frac{1}{V}\left(\frac{\partial V}{\partial P}\right)_T \quad \text{and} \quad \kappa_S = -\frac{1}{V}\left(\frac{\partial V}{\partial P}\right)_S$$

We have

$$\frac{C_P}{C_V} = \frac{T\left(\frac{\partial S}{\partial T}\right)_P}{T\left(\frac{\partial S}{\partial T}\right)_V} = \frac{\left(\frac{\partial S}{\partial T}\right)_P}{\left(\frac{\partial S}{\partial T}\right)_V} \qquad [7.15]$$

Now we cannot eliminate S directly in equation [7.15] by using the Maxwell relations, as recourse to our mnemonic immediately shows. Indeed, we do not wish to do so, as we need an S outside the partial differential to give κ_S. Let us then recast the order in the partial differentials in equation [7.15] by using the cyclical rule

$$\left(\frac{\partial S}{\partial T}\right)_P \left(\frac{\partial P}{\partial S}\right)_T \left(\frac{\partial T}{\partial P}\right)_S = -1$$

so $\quad \left(\frac{\partial S}{\partial T}\right)_P = -\left(\frac{\partial S}{\partial P}\right)_T \left(\frac{\partial P}{\partial T}\right)_S$ [7.16]

and similarly,

$$\left(\frac{\partial S}{\partial T}\right)_V = -\left(\frac{\partial S}{\partial V}\right)_T \left(\frac{\partial V}{\partial T}\right)_S$$ [7.17]

Substituting equations [7.16] and [7.17] in equation [7.15],

$$\frac{C_P}{C_V} = \frac{\left(\frac{\partial S}{\partial P}\right)_T \left(\frac{\partial P}{\partial T}\right)_S}{\left(\frac{\partial S}{\partial V}\right)_T \left(\frac{\partial V}{\partial T}\right)_S} = \frac{\left(\frac{\partial S}{\partial P}\right)_T \left(\frac{\partial V}{\partial S}\right)_T}{\left(\frac{\partial V}{\partial T}\right)_S \left(\frac{\partial T}{\partial P}\right)_S}$$

$$= \frac{\left(\frac{\partial V}{\partial P}\right)_T}{\left(\frac{\partial V}{\partial P}\right)_S}$$

So $\quad \boxed{\dfrac{C_P}{C_V} = \dfrac{\kappa_T}{\kappa_S}}$ [7.18]

This is again, at first sight, a surprising result, relating elasticity parameters to heat capacities, but it should be realised that the C's are determined under definite conditions of P and V while the κ's are determined under conditions of definite T and S. Hence, at second sight, it is not really so surprising that results such as equations [7.18] and [7.6] exist.

As an application of equation [7.18], let us move from a fluid system to a weakly magnetic system, where the infinitesimal work term is given by $B_0 d \cdot \mathcal{M}$ rather than $- P \, dV$ and where magnetic work is the only work. V has then to be replaced by $\cdot \mathcal{M}$ and P by $- B_0$. The corresponding relation to equation [7.18] is

$$\frac{C_{B_0}}{C_{\mathcal{M}}} = \frac{\chi_T}{\chi_S} \qquad [7.19]$$

where the isothermal and adiabatic differential susceptibilities are:

$$\chi_T = \frac{\mu_0}{V} \left(\frac{\partial \cdot \mathcal{M}}{\partial B_0} \right)_T \quad \text{and} \quad \chi_S = \frac{\mu_0}{V} \left(\frac{\partial \cdot \mathcal{M}}{\partial B_0} \right)_S \qquad [7.20]$$

Thus, by measuring the ratio of the magnetic susceptibilities, which are relatively easy to measure, the ratio of the magnetic specific heats may be determined.

In a paramagnetic salt, the magnetism is due to unpaired effective electron spins. The best known paramagnetic salt, because of its use as a thermometer below 1 K, is cerium magnesium nitrate, $Ce_2Mg_3(NO_3)_{12} \cdot 24H_2O$, where the single unpaired spin is on the cerium ion. We may regard the cerium spins as the system while the magnesium nitrate lattice can be regarded as the surroundings, as shown in Fig. 7.1(a). Susceptibilities may be measured using the AC method shown in Fig. 7.1(b).

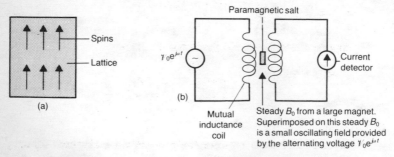

Fig. 7.1 The relaxation method of measuring the ratio of magnetic heat capacities.

The spins are aligned with a steady magnetic field and, super-imposed on this, is a small AC field which flips the spins parallel and antiparallel to the steady field. The current measured by the detector is proportional to the susceptibility χ of the sample.

Now the spins can flip over in a characteristic time τ, giving up their excess magnetic energy to the lattice. For cerium magnesium nitrate, this time is ~ 0.01 s at 2 K. At low measuring frequencies, with a period much longer than τ, the spins have plenty of time to exchange their energy of magnetisation with the lattice, and we measure χ_T. Conversely, at high frequencies, we measure χ_S. From these measurements of χ_T and χ_S we can determine the ratio $C_{B_0}/C_{\mathcal{M}}$.

This technique is known as the *relaxation method of measuring heat capacities* and is important in low temperature physics.

7.5 The entropy of an ideal gas again

Now that we have more powerful techniques at our disposal, it is very instructive to re-derive equation [5.11] which gives the entropy of an ideal gas as a function of temperature and volume. We write

$$S = S(T, V) \tag{7.21}$$

so
$$\mathrm{d}S = \left(\frac{\partial S}{\partial T}\right)_V \mathrm{d}T + \left(\frac{\partial S}{\partial V}\right)_T \mathrm{d}V \tag{7.22}$$

or
$$\mathrm{d}S = C_V \frac{\mathrm{d}T}{T} + \left(\frac{\partial P}{\partial T}\right)_V \mathrm{d}V \tag{7.23}$$

using equation [6.9] and the Maxwell relation

$$\left(\frac{\partial S}{\partial V}\right)_T = \left(\frac{\partial P}{\partial T}\right)_V \tag{6.26}$$

Let us consider one mole of the gas with the molar heat capacity at a constant volume c_v. Then

$$Pv = RT$$

and $\left(\dfrac{\partial P}{\partial T}\right)_v = \dfrac{R}{v}$

Hence equation [7.23] becomes

$$ds = c_v \frac{dT}{T} + R \frac{dv}{v}$$

Integrating,

$$\boxed{s = c_v \ln T + R \ln v + s_0} \qquad [5.11]$$

where s_0 is a constant. For n moles, this becomes

$$\boxed{S = n(c_v \ln T + R \ln v + s_0)}$$

We may, in exactly the same way, find the entropy as a function of temperature and pressure. We write

$$S = S(T, P) \qquad [7.24]$$

so $\quad dS = \left(\dfrac{\partial S}{\partial T}\right)_P dT + \left(\dfrac{\partial S}{\partial P}\right)_T dP \qquad [7.25]$

But $C_P = T(\partial S / \partial T)_P$ and we have the Maxwell relation

$$\left(\frac{\partial S}{\partial P}\right)_T = -\left(\frac{\partial V}{\partial T}\right)_P \qquad [6.41]$$

Hence equation [7.25] becomes

$$dS = C_P \frac{dT}{T} - \left(\frac{\partial V}{\partial T}\right)_P dP \qquad [7.26]$$

If we now consider one mole, this becomes

$$dS = c_P \frac{dT}{T} - R \frac{dP}{P} \qquad [7.27]$$

using the equation of state for one mole. Integrating equation [7.27],

$$\boxed{s = c_P \ln T - R \ln P + s_0}$$ [7.28]

for one mole or

$$\boxed{S = n(c_P \ln T - R \ln P + s_0)}$$

for n moles.

7.6 The Joule coefficient for a free expansion

Let us return to the calculation of the temperature change in a non-ideal gas when it undergoes a free expansion as shown in Fig. 2.7. We saw in section 3.6 that no work is done, no heat enters the system as the walls are adiabatic and the internal energy is unchanged in the expansion. For an ideal gas, this means that there is no temperature change.

We note first that a free expansion is an *irreversible* process, with the gas *not* passing through a series of equilibrium states. However the *end points are equilibrium states*, and it is for this reason that we may apply thermodynamics to this process to determine the temperature change.

If the subscripts i and f refer as usual to the initial and final equilibrium points, we know that $U_i = U_f$. Now T_i and T_f are uniquely specified by the pairs of variables (U_i, V_i) and (U_f, V_f) and so $\Delta T = T_f - T_i$ is also uniquely specified. Hence, let us *imagine* a reversible expansion from the state i to the state f and calculate ΔT for this. The ΔT so obtained will be the same as the actual ΔT occurring in the irreversible expansion. The most convenient reversible process connecting the end points is a quasistatic expansion occurring at constant U.

In the light of our comment in the middle of section 2.10, we write

$$T = T(V, U)$$

so $\qquad \mathrm{d}T = \left(\frac{\partial T}{\partial V}\right)_U \mathrm{d}V + \left(\frac{\partial T}{\partial U}\right)_V \mathrm{d}U$ [7.29]

where the second term on the right is zero as we are considering a constant U process. Integrating,

$$\Delta T = \int_{V_i}^{V_f} \left(\frac{\partial T}{\partial V} \right)_U dV \qquad [7.30]$$

The partial differential $(\partial T / \partial V)_U$ is the Joule coefficient μ_J and, in order to be able to integrate equation [7.30], we have to find an expression for μ_J in terms of P, V and T. Let us now do this.

The first thing to recognise about μ_J is that it is difficult to handle as it stands because of the constant U outside the partial differential, although we are perfectly happy with U inside a partial differential, as in $C_V = (\partial U / \partial T)_V$ and in the energy equation for example. We can bring U inside using the cyclical rule:

$$\left(\frac{\partial T}{\partial V} \right)_U \left(\frac{\partial U}{\partial T} \right)_V \left(\frac{\partial V}{\partial U} \right)_T = -1$$

or $\quad \left(\frac{\partial T}{\partial V} \right)_U = - \left(\frac{\partial T}{\partial U} \right)_V \left(\frac{\partial U}{\partial V} \right)_T \qquad [7.31]$

so $\quad \mu_J = -\frac{1}{C_V} \left(\frac{\partial U}{\partial V} \right)_T \qquad [7.32]$

as $\quad C_V = \left(\frac{\partial U}{\partial T} \right)_V \qquad [6.8]$

But the partial differential in equation [7.32] is given by the energy equation:

$$\left(\frac{\partial U}{\partial V} \right)_T = T \left(\frac{\partial P}{\partial T} \right)_V - P \qquad [7.10]$$

So $\quad \boxed{\mu_J = \frac{1}{C_V} \left\{ P - T \left(\frac{\partial P}{\partial T} \right)_V \right\}} \qquad [7.33]$

The Joule coefficient may be calculated, using equation [7.33] and the equation of state. Let us consider two examples of the calculation of μ_J.

The Joule coefficient for an ideal gas

We have, for one mole of an ideal gas,

$$Pv = RT$$

so $$\left(\frac{\partial P}{\partial T}\right)_v = R/v$$

Substituting this in equation [7.33] gives $\mu_J = 0$, as expected, with no resulting temperature change.

The Joule coefficient for a real gas

A useful modification to $Pv = RT$ for one mole of a real gas is the so called *virial expansion*:

$$Pv = RT(1 + B_2/v + B_3/v^2 + \ldots) \qquad [7.34]$$

where the B_n's are the virial coefficients which are temperature dependent and get progressively smaller the higher the order of the term. Let us go only as far as the B_2 term and neglect the smaller higher-order terms. Then, if we substitute equation [7.34] in equation [7.33], we find, after a little manipulation, that

$$\mu_J = -\frac{1}{c_v}\frac{RT^2}{v^2}\frac{dB_2}{dT} \qquad [7.35]$$

The variation of B_2 with temperature is known. For argon, for example, where $dB_2/dT = 0.25$ cm^{3}mol^{-1}K^{-1}, μ_J can be calculated from equation [7.35] to be:

$$\mu_J = -2.5 \times 10^{-5} \text{ K mol cm}^{-3}$$

Suppose now we double the volume of one mole of argon at STP. We remember that one mole of a gas at STP occupies 22.4 l. In order to gain an order of magnitude answer, let us assume that μ_J is constant for the integration in equation [7.30]. Then

$$\Delta T \approx \mu_J \Delta v$$
$$\approx -2.5 \times 10^{-5} \times 22.4 \times 10^3 = -0.6 \text{ K}$$

which is a very small effect. Indeed, as has been mentioned in section 3.6, Joule was unable to measure this effect for air.

The free expansion of any gas *always* results in cooling. The physical reason for this was discussed earlier in section 3.6.

7.7 The Joule–Kelvin coefficient for the throttling process

We conclude this chapter with a discussion of the important Joule–Kelvin effect or throttling process. This process is illustrated in Fig. 3.5. It was also seen in section 3.8 that the process takes place with there being no change in the enthalpy. Before we enter into a discussion of the thermodynamics of this irreversible process, let us describe the effect in a little more detail.

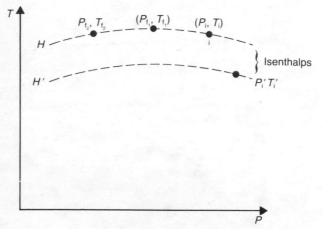

Fig. 7.2 The isenthalps of a throttling process.

The most convenient way to represent the effect is on a temperature-versus-pressure plot, as in Fig. 7.2. Suppose that the initial equilibrium state before throttling is the point i at (P_i, T_i). With the gas in this initial state, we could throttle the gas to the lower pressure P_{f_1} and we could also measure the final temperature T_{f_1} so that the gas ends up in the final equilibrium state (P_{f_1}, T_{f_1}). This state will have the same enthalpy as i. If now we were to change the final pressure to a lower value P_{f_2}, but were to throttle the gas from the same initial state i, the gas would end in a new final equilibrium state (P_{f_2}, T_{f_2}) with the same enthalpy as i again. By repeating this experiment many times, we could obtain a series of points representing the different final equilibrium points, all starting from the same initial equilibrium point and all with the same enthalpy as i. The curve joining them is called an *isenthalp*. However, it must be emphasised very strongly that an isenthalp is not a curve representing *one* given throttling process between the equilibrium points at the ends of the isenthalp; instead, it is the locus of the end points of *different* throttling processes, all starting at the same initial state.

If now a second initial point (P_i', T_i') is chosen, a second isenthalp may be drawn and this is shown as the lower curve in Fig. 7.2.

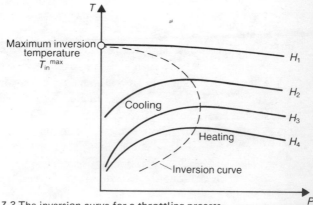

Fig. 7.3 The inversion curve for a throttling process.

We can then draw a whole series of experimentally determined isenthalps and the results would appear as in Fig. 7.3 for a typical gas. The maxima of the different isenthalps lie on the *inversion curve*. Depending on where the initial and final points are chosen, both heating and cooling effects can be produced, in contrast to the free expansion, where there is always cooling. The inversion curve separates the region of heating from the region of cooling. The greatest cooling effect, for a given pressure drop and starting temperature, will occur when the initial state i is on the inversion curve. The temperature coordinate of a point on the inversion curve is called the inversion temperature T_{in}. For there to be a cooling effect, the initial temperature must be chosen to be less than the *maximum inversion temperature* T_{in}^{max}, shown in Fig. 7.3, which is at the intersection of the inversion curve and the temperature axis. The maximum inversion temperatures for some common gases are given in Table 7.1.

Table 7.1 *Maximum inversion temperatures for some common gases*

Gas	Maximum inversion temperature $T_{in}\,^{max}$
Argon	723 K
Nitrogen	621 K
Hydrogen	205 K
Helium	51 K

You are asked to show in question 15, Chapter 7, Appendix 4, that the maximum inversion temperature for a van der Waals gas is given approximately by $2a/Rb$.

Let us now derive an expression for ΔT for a throttling process in terms of P, V and T. As for the free expansion, we recognise that, since the end points at (P_i, H_i) and (P_f, H_f) are equilibrium states, the temperature change calculated for an *imaginary reversible* process between them is the same as the temperature change in the *actual irreversible* throttling process. The most convenient reversible process to choose is a quasistatic expansion from the initial pressure of P_i to the final pressure of P_f taking place at constant enthalpy. Thus we write

$$T = T(P,H)$$

so that

$$dT = \left(\frac{\partial T}{\partial P}\right)_H dP + \left(\frac{\partial T}{\partial H}\right)_P dH \qquad [7.36]$$

where the second term on the right vanishes as we are considering a constant H process. Integrating,

$$\Delta T = \int_{P_i}^{P_f} \left(\frac{\partial T}{\partial P}\right)_H dP \qquad [7.37]$$

Lancaster University Library

Items Issued Subject To Recall

Title: 64008390U
ID: 64008390U
Due: 09/03/2013 12:00 AM

Total items: 1
23/02/2013 07:34 PM

Important - the fine for the late return of
recalled books will increase to £2.00 per day
from 1 August 2012. Avoid fines by checking
your emails regularly for recall notices.

The partial differential $(\partial T/\partial P)_H$ is the Joule–Kelvin coefficient μ_{JK} and, in order to effect the integration in equation [7.37], we have to express μ_{JK} in terms of P, V and T. Let us now do this.

We immediately recognise that the difficulty with

$$\mu_{JK} = \left(\frac{\partial T}{\partial P}\right)_H \qquad [7.38]$$

is the constant H outside the partial differential. It can be brought inside using the cyclical rule and hopefully we can combine it with $\mathrm{d}T$ as $(\partial H/\partial T)_P$, which is C_P. Applying the cyclical rule,

$$\left(\frac{\partial T}{\partial P}\right)_H \left(\frac{\partial H}{\partial T}\right)_P \left(\frac{\partial P}{\partial H}\right)_T = -1$$

or $\quad \left(\dfrac{\partial T}{\partial P}\right)_H = -\left(\dfrac{\partial T}{\partial H}\right)_P \left(\dfrac{\partial H}{\partial P}\right)_T \qquad [7.39]$

The first term on the right of equation [7.39] is indeed $1/C_P$, while the second is the enthalpy counterpart of the energy equation for U. Let us now find $(\partial H/\partial P)_T$.

We have our basic definition

$$H = U + PV \qquad [3.7]$$

giving as before

$$\mathrm{d}H = T\,\mathrm{d}S + V\,\mathrm{d}P \qquad [6.11]$$

so $\quad \left(\dfrac{\partial H}{\partial P}\right)_T = T\left(\dfrac{\partial S}{\partial P}\right)_T + V \qquad [7.40]$

But we have the Maxwell relation

$$\begin{array}{c} S \\ \overline{} \\ P \qquad V \\ T \end{array} \qquad \left(\frac{\partial S}{\partial P}\right)_T = -\left(\frac{\partial V}{\partial T}\right)_P \qquad [6.41]$$

Substituting equation [6.41] into equation [7.40]

$$\left(\frac{\partial H}{\partial P}\right)_T = V - T\left(\frac{\partial V}{\partial T}\right)_P \qquad [7.41]$$

Hence equation [7.39] becomes

$$\mu_{JK} = \frac{1}{C_P} \left\{ T \left(\frac{\partial V}{\partial T} \right)_P - V \right\} \qquad [7.42]$$

which can be calculated from the equation of state and may be positive or negative. Finally, the temperature change in the throttling process may be determined by substituting the value of μ_{JK} given by equation [7.42] into equation [7.37].

It is easily shown that μ_{JK} is zero for an ideal gas and so there is no temperature change when an ideal gas undergoes a throttling process. For a real gas, μ_{JK} is best calculated from equation [7.42] by writing the equation of state in a virial form such as equation [7.34]. All the virial coefficients B_i and their temperature coefficients are tabulated in the reference handbooks such as Kaye and Laby (see Appendix 6).

The throttling process is of enormous importance, especially in the liquefaction of gases. We shall see an application of this process in the next and final section of this chapter when we consider the classic Linde method of gas liquefaction.

7.8 The Linde liquefaction process

A schematic representation of the Linde liquefier is given in Fig. 7.4. The most difficult gas to liquefy is helium as it does not do

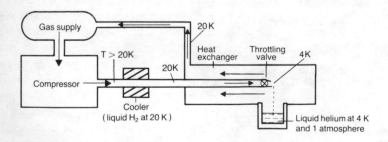

Fig. 7.4 A schematic representation of a Linde liquefier.

so until the temperature has been lowered to 4 K (at a pressure of 1 atmosphere). We shall discuss the use of the Linde liquefier for helium, but of course it can be used for other gases.

In order for there to be any cooling using the throttling process, the helium gas has first to be cooled below the inversion temperature of 51 K by passing it through the cooler, which is a coiled pipe immersed in a bath of liquid hydrogen at 20 K. The gas then enters the countercurrent heat exchanger at 20 K, at a high pressure, where it is throttled through a valve. There it undergoes cooling, but not enough to cause liquefaction immediately. The cooled gas passes out through the heat exchanger and, in doing so, cools the incoming gas below 20 K. This gas in its turn expands and cools the next amount of incoming gas even more. Eventually, the temperature on the inlet side of the throttling valve is low enough for liquefaction to occur and the liquid helium collects at the bottom of the heat exchanger container at 4 K and at a pressure of 1 atmosphere.

The compressor drives the helium gas around the circuit and provides the necessary high pressure at the inlet side of the throttling valve. After compression and the consequent heating, the gas is cooled back to 20 K again by the liquid hydrogen cooler so that the helium always enters the heat exchanger at this temperature. It must be understood that the liquefier does *not* work on the principle that a given mass of gas completes several circuits suffering successive and additive temperature drops until it eventually liquefies.

Although Linde liquefiers work very well for helium, they have the severe disadvantage that they are dangerous because of the use of liquid hydrogen. This substance is extremely explosive and the number of hydrogen fires and explosions in low-temperature physics laboratories are too numerous to list. One of the most common helium liquefiers currently in use in many Western low-temperature laboratories is the American Collins machine. In this liquefier, the helium gas in pre-cooled to 77 K with liquid nitrogen; it is then cooled to below its inversion temperature by allowing it to perform adiabatic external work in a successive pair of expansion engines. Finally it is liquefied with a throttling process.

Chapter 8

Magnetic systems, radiation, rubber bands and electrolytic cells

So far in this book we have confined our attention to systems in which the state variables are P, V and T, which are related by an equation of state so that only two of them are independent. In this chapter we turn our attention to other systems which are described by different sets of state variables. Although our choice could be considerably wider, because of limitations on space we have chosen to discuss only three such systems: a magnetic system; a rubber band; and a reversible electrolytic cell. We shall also discuss a fourth system which, although it is described in terms of P, V and T, differs sufficiently in nature from our familiar fluid PVT systems to warrant inclusion here. This fourth system is thermal radiation in a cavity.

These four systems will serve to illustrate the wide applicability of the methods of thermodynamics.

MAGNETIC SYSTEMS

8.1 The first law for a magnetic system

In our discussion of magnetic systems we shall use the SI (Sommerfeld) system of units. Unfortunately, there is some confusion over units in the field of magnetism and we follow the *clear* exposition on these matters in the book by Crangle (see Appendix 6).

In the SI system, the following relation holds:

$$B = \mu_0(H + M) \qquad [8.1]$$

where B is the magnetic induction, H is the applied magnetising field or, more commonly, the magnetic field, and M is the magnetisation, or the magnetic moment per unit volume. Although all these quantities are vectors, we shall assume that they are all parallel and so treat them as scalars.

H is a convenient field because we can relate it to the currents in wires I_f producing it using the standard relation $\oint \mathbf{H.ds} = I_f$ (Dobbs, 1984, equation [6.10]) but B is the more fundamental field from a microscopic point of view because the Lorentz force experienced by the moving electrons in the magnetic atoms is $e \, \mathbf{v} \times \mathbf{B}$ (Dobbs, 1984, equation [4.8]) where e is the electronic charge and $\mathbf{v}$ the velocity. We shall thus develop our discussion in terms of B rather than H. Another reason for this choice is that *magnetic fields are virtually always quoted in tesla*, the unit of induction.

The induction B in a material is composed of two parts; the induction $\mu_0 H$ that would be present in free space in the absence of the material and which we shall call B_0; and the contribution $\mu_0 M$ from the material arising from the net circulating currents in the elementary atomic magnets. We write then

$$B = B_0 + \mu_0 M \qquad [8.2]$$

The total magnetic moment $\mathcal{M}$ of the sample, assuming that the magnetisation is uniform throughout the sample as it is for most situations of importance, is

$$\mathcal{M} = M V \qquad [8.3]$$

It is found, for many materials, that there is a unique dependence of $\mathcal{M}$ on T and B_0, so that

$$\mathcal{M} = \mathcal{M}(B_0, T) \qquad [8.4]$$

Because of hysteresis, such a relation does *not* hold for ferromagnetic materials, and we have to exclude such materials from our discussion. It is further found that, for many materials, the magnetisation M is proportional to B_0. We define the measured or bulk magnetic susceptibility χ_m in terms of the applied fields H and B_0 as

$$\chi_m = M/H = \mu_0 M/B_0 = \frac{\mu_0 \mathcal{M}}{V B_0} \qquad [8.5]$$

For such linear materials, equations [8.1] and [8.2] become

$$B = \mu_0(1 + \chi_m) H = (1 + \chi_m)B_0 \qquad [8.6]$$

We shall restrict ourselves to the special case of magnetically weak materials where $\chi_m \ll 1$, so that

$$B = B_0 \qquad [8.7]$$

The value of χ_m for the paramagnetic salts of interest in our subsequent discussion is only $\sim 10^{-2}$ or 10^{-3} while diamagnetic materials have values of $\sim 10^{-6}$, so this approximation is justified. (For our magnetically weak systems we may ignore the thorny complication of demagnetising fields which result in the measured susceptibility being dependent on the overall sample shape. These effects are discussed in the standard texts on magnetism such as Crangle (see Appendix 6).)

Our interest will centre on paramagnetic materials, many of which are found to obey the Curie law

$$\chi_m = \mathscr{C}/T \qquad [8.8]$$

down to very low temperatures. For the purpose of our discussion here we shall take this law, which relates $\mathscr{M}$, B_0 and T, as the equation of state for our magnetic system. The constant $\mathscr{C}$ is called the Curie constant. The Curie-Weiss law

$$\chi_m = \frac{\mathscr{C}}{T - T_0} \qquad [8.9]$$

is a more general modifcation of the Curie law which holds for samples that are not magnetically weak and where the interaction between the magnetic ions is important. T_0 is the Curie-Weiss constant which is usually only a fraction of a degree for the paramagnetic salts that we shall consider, so the Curie law is a good approximation.

As is shown in Appendix 3, the appropriate form for the infinitesimal work term in a magnetic system is

$$dW = B_0 d . \mathscr{M} \qquad [8.10]$$

Thus the infinitesimal form of the first law is

$$dU = dQ - P \, dV + B_0 d . \mathscr{M} \qquad [8.11]$$

Let us concern ourselves with situations in which only the applied induction field B_0 is changed, there being no change in the pressure. Then, in practice, we may ignore the $-P\mathrm{d}V$ term compared with the $B_0\mathrm{d}.\mathscr{M}$ term, as the second term is so much larger than the first. The change in volume of a magnetic system upon the application of an external induction field is known as *magnetostriction*. This effect is always small and is certainly negligible for the paramagnetic salts that we shall consider. We may then write the first law as

$$\mathrm{d}Q = \mathrm{d}U - B_0\mathrm{d}.\mathscr{M} \qquad [8.12]$$

The thermodynamic treatment of magnetic systems follows that for our familiar PVT system, with P being replaced by $-B_0$ and V by $.\mathscr{M}$. The four invaluable Maxwell relations can be obtained from the modified mnemonic:

$$
\begin{array}{ccc}
 & S & \\
\overline{} & & \\
-B_0 & & .\mathscr{M} \\
 & T &
\end{array}
$$

The extra minus sign on B_0 should not cause difficulty if it is realised that a constant $-B_0$ outside a partial differential is the same as a constant $+B_0$. The reader should try to produce, for example, the Maxwell relation

$$\left(\frac{\partial S}{\partial B_0}\right)_T = \left(\frac{\partial.\mathscr{M}}{\partial T}\right)_{B_0}$$

both from this mnemonic *and also* using the full analysis involving the Gibbs function, G.

8.2 Magnetic cooling

One of the currently exciting frontiers of physics is the temperature region close to the absolute zero. Although the technique of magnetic cooling was first used in the 1930s, it is still the standard technique used to achieve the very lowest temperatures. The recent introduction of the ^{3}He-^{4}He dilution refrigerator has meant that temperatures down to 4×10^{-3} K can be maintained, but temperatures lower than this have to be produced using the technique of magnetic cooling. It is possible to produce temperatures as low as 10^{-6} K using this method in which the elementary magnetic dipoles comprising the magnetic system are the *nuclear* spins of copper. However, we shall discuss the application of this technique in cooling a set of *electron* spins in a paramagnetic salt. Before we enter into a detailed thermodynamic analysis, let us first discuss the principle of the method.

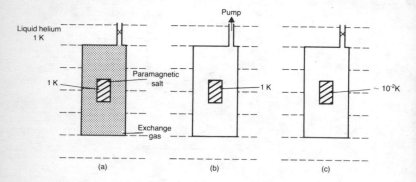

Fig. 8.1 The different stages used in adiabatic demagnetisation.

In Fig. 8.1 we show the different stages used in this technique for cooling a paramagnetic salt from a starting temperature of 1 K. The salt is suspended by fine cotton threads in the middle of a chamber immersed in a bath of liquid helium at 1 K. Surrounding the salt is exchange gas, again helium, which may be pumped away so that the salt may be thermally isolated from the surrounding helium bath. The sequence of operations is as follows.

1. The salt is magnetised with the application of a large magnetic induction B_0 of the order of one or more tesla. This induction field is provided by an electromagnet or a superconducting magnet; the latter can produce enormous fields of 10 tesla or more. Because of the presence of the exchange gas, the magnetisation is isothermal. The heat of magnetisation is conducted away to the helium bath by the exchange gas.

2. The exchange gas is pumped away so that the salt is thermally isolated.

3. The applied induction is slowly reduced to zero so that the salt is always in a state of thermodynamic equilibrium and the demagnetisation proceeds reversibly. As the process is also adiabatic, the demagnetisation is thus *isentropic*. The temperature is then found to fall dramatically. For a starting temperature of 1 K, a demagnetisation temperature of $\sim 1/100$ K is typical.

To understand the physical reason for this effect, we have to consider the entropy curves shown in Fig. 8.2(a). The upper curve

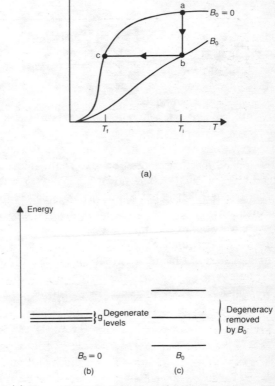

(a)

(b)   (c)

Fig. 8.2 (a) The entropy as a function of temperature for a paramagnetic salt in an applied induction field B_0 and in zero field. The process ab represents an isothermal magnetisation while the process bc represents an adiabatic demagnetisation to the temperature T_f. (b) and (c) The g-fold degeneracy of the energy levels is removed upon the application of B_0.

shows the entropy of the salt in the absence of an applied magnetic induction. At temperatures of the order of 1 K, only the g-fold degenerate ground energy level of the salt is occupied. The degeneracy g is usually a small number such as 2 or 3. This is illustrated in Fig. 8.2(b). Each magnetic atom has g possible ways of entering the ground level and so the number of different ways of arranging the N atoms is $\Omega = g^N$. The entropy of the magnetic system is then, using equation [5.12],

$$S = k_B \ln g^N$$

or $\quad S = k_B N \ln g$

In fact, the weak magnetic coupling between the neighbouring electron spins splits the states of the ground level so that they are not actually degenerate. This splitting is so small that, at ~ 1 K, all these states are thermally occupied and the expression for the entropy just obtained holds. However, at lower temperatures only the actual ground state is occupied with $\Omega = 1$ and the entropy falls to zero, as shown in the figure.

The lower curve of Fig. 8.2(a) shows the entropy in an externally applied magnetic induction B_0. The application of such a field splits the ground level as shown in Fig. 8.2(c). The degeneracy is then removed so that the entropy is reduced. An alternative viewpoint is to note that the application of B_0 aligns the dipoles, thus imposing more order on the spin system and reducing the entropy. It is a consequence of the third law of thermodynamics, to be discussed in Chapter 11, that the two entropy curves meet at absolute zero.

We can now see the physical reason for the drop in temperature. The isothermal magnetisation is represented by the path ab in Fig. 8.2(a). The isentropic adiabatic demagnetisation is represented by the path bc. As long as the demagnetisation process takes place in the region of the shoulder of the entropy curve, significant cooling occurs. To achieve temperatures close to the absolute zero, the magnetic salt should be chosen so that this shoulder is at very low temperatures.

Some typical paramagnetic salts used are:
1. *iron* ammonium alum;

2. *gadolinium* sulphate;

3. *cerium* magnesium nitrate.

The magnetic ion, containing a number of unpaired electron spins, has been written in italics in each case. It might be asked why one uses such apparently obscure compounds. The answer is that the non-magnetic part of the compound acts as a dilutant, keeping the magnetic atoms well separated and so reducing the interaction between them. This ensures that the fall off in the entropy due to this interaction does not occur until very low temperatures, as required.

There is in fact another reason for working within the vicinity of the shoulder. It would be absolutely pointless to demagnetise a magnetic salt if it immediately warmed up again because of the inevitable 'heat leak' into the system from the outside. One can never completely thermally isolate the salt. Fortunately, the steepness of the shoulder of the entropy curve in exactly the region in which we are working ensures an *enormous* heat capacity $(C = T \, dS/dT)$ to act as a thermal ballast. Indeed, the heat capacity of 1 cm^3 of iron ammonium alum at 0.01 K is equal to that of 16 tons of lead! This ensures that the salt in practice remains cold for a sufficiently long time, perhaps in the region of hours, for sensible measurements to be made.

8.3 Thermodynamic analysis of magnetic cooling

Let us now calculate the cooling produced in the adiabatic demagnetisation bc shown in Fig. 8.2. The calculation of the heat produced in the isothermal magnetisation ab is left as question 1, Chapter 8, Appendix 4.

As we are told the changes in the state functions B_0 and S, we write

$$T = T(B_0, S) \tag{8.13}$$

so
$$dT = \left(\frac{\partial T}{\partial B_0}\right)_S dB_0 + \left(\frac{\partial T}{\partial S}\right)_{B_0} dS$$

or
$$dT = \left(\frac{\partial T}{\partial B_0}\right)_S dB_0 \tag{8.14}$$

as dS is zero.

We can bring the S appearing in equation [8.14] inside the partial differential, with the intention of coupling it with a dT to produce a heat capacity, using the cyclical rule:

$$\left(\frac{\partial T}{\partial B_0}\right)_S \left(\frac{\partial S}{\partial T}\right)_{B_0} \left(\frac{\partial B_0}{\partial S}\right)_T = -1 \qquad [8.15]$$

or $\quad \left(\frac{\partial T}{\partial B_0}\right)_S = -\left(\frac{\partial T}{\partial S}\right)_{B_0} \left(\frac{\partial S}{\partial B_0}\right)_T \qquad [8.16]$

Thus $\left(\frac{\partial T}{\partial B_0}\right)_S = -\frac{T}{C_{B_0}} \left(\frac{\partial \mathcal{M}}{\partial T}\right)_{B_0} \qquad [8.17]$

as $C_{B_0} = T(\partial S/\partial T)_{B_0}$ and the Maxwell relation $(\partial \mathcal{M}/\partial T)_{B_0} = (\partial S/\partial B_0)_T$ holds. We have

$$\chi_m = \frac{\mu_0 M}{B_0} = \frac{\mu_0 \cdot \mathcal{M}}{V B_0} \qquad [8.5]$$

so $\quad \frac{V B_0}{\mu_0} \left(\frac{\partial \chi_m}{\partial T}\right)_{B_0} = \left(\frac{\partial \mathcal{M}}{\partial T}\right)_{B_0} \qquad [8.18]$

Substituting equation [8.18] into equation [8.17],

$$\left(\frac{\partial T}{\partial B_0}\right)_S = -\frac{T V B_0}{C_{B_0} \mu_0} \left(\frac{\partial^2 \chi_m}{\partial T}\right)_{B_0} \qquad [8.19]$$

where we know $(\partial \chi_m/\partial T)_{B_0} = -\mathcal{C}/T^2$ from the Curie law.

This is essentially the end of our argument, as we can now substitute equation [8.19] into equation [8.14] and integrate with respect to B_0 and T to obtain the temperature fall. However, before we can do this, we have to find the B_0 dependence of $C_{B_0}(T, B_0)$. By the B_0 dependence of C_{B_0} we mean that C_{B_0}, the heat capacity in *constant* magnetic induction, may have different values when determined in *different* but *steady* B_0's. Let us now find this dependence.

In exact analogy to equation [7.9] we have

$$\left(\frac{\partial C_{B_0}}{\partial B_0}\right)_T = T\left(\frac{\partial^2 \cdot \mathcal{M}}{\partial T^2}\right)_{B_0} \qquad [8.20]$$

or
$$\left(\frac{\partial C_{B_0}}{\partial B_0}\right)_T = \frac{TVB_0}{\mu_0}\left(\frac{\partial^2 \chi_m}{\partial T^2}\right)_{B_0}$$ [8.21]

using equation [8.5]. As well as obeying the Curie law, paramagnetic salts usually have a so-called *Schottky* temperature dependence of the heat capacity C_{B_0} in zero magnetic induction given by

$$C_{B_0}(T,0) = \frac{Vb}{T^2}$$ [8.22]

where b is a constant. Using equation [8.8] in equation [8.21]

$$\left(\frac{\partial C_{B_0}}{\partial B_0}\right)_T = \frac{TVB_0}{\mu_0}\frac{2\mathscr{C}}{T^3}$$ [8.23]

Integrating at constant temperature,

$$C_{B_0}(T,B_0) = \frac{V\mathscr{C}B_0^2}{\mu_0 T^2} + C_{B_0}(T,0)$$

or
$$\boxed{C_{B_0}(T,B_0) = \frac{V}{T^2}\left(b + \frac{\mathscr{C}B_0^2}{\mu_0}\right)}$$ [8.24]

using equation [8.22]. We may now substitute equation [8.19] into equation [8.14] using the expression for C_{B_0} given in equation [8.24]:

$$\mathrm{d}T = \left(\frac{\partial T}{\partial B_0}\right)_S \mathrm{d}B_0 = -\frac{TVB_0}{\mu_0}\frac{1}{\dfrac{V}{T^2}\left(b + \dfrac{\mathscr{C}B_0^2}{\mu_0}\right)}$$

$$\times \left(\frac{-\mathscr{C}}{T^2}\right)\mathrm{d}B_0$$ [8.25]

Thus,

$$\frac{\mathrm{d}T}{T} = \frac{\mathscr{C}B_0\mathrm{d}B_0}{\mu_0\left(b + \dfrac{\mathscr{C}B_0^2}{\mu_0}\right)}$$ [8.26]

Integrating,

$$\ln T \Big|_i^f = \frac{1}{2} \ln \left(b + \frac{\mathscr{C}B_0^2}{\mu_0} \right) \Big|_i^f \qquad [8.27]$$

or
$$\frac{T_f}{T_i} = \left(\frac{b + \dfrac{\mathscr{C}B_{0f}^2}{\mu_0}}{b + \dfrac{\mathscr{C}B_{0i}^2}{\mu_0}} \right)^{\frac{1}{2}} \qquad [8.28]$$

where i and f refer to the initial and final conditions. If the salt is demagnetised to zero applied induction, the final temperature reached is:

$$\boxed{T_f = T_i \left(\frac{b}{b + \dfrac{\mathscr{C}B_{0i}^2}{\mu_0}} \right)^{\frac{1}{2}}} \qquad [8.29]$$

We expect this relation to hold for a salt obeying the Curie law and where the heat capacity in zero induction is given by equation [8.22]. Gadolinium sulphate is one such salt. In Fig. 8.3 we present the experimental results. The linear dependence of $(T_i/T_f)^2$ on B_{0i}^2 is apparent.

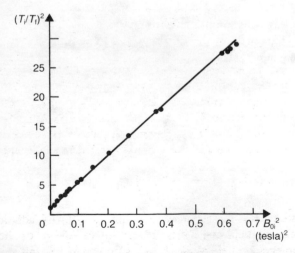

Fig. 8.3 The linear dependence of $(T_i/T_f)^2$ with B_{0i}^2 in agreement with equation [8.29]. The measurements are for gadolinium sulphate.

8.4 Radiation as a *PVT* system

We are all familiar with the fact that a hot poker glows dull red. In fact, all bodies emit electromagnetic radiation by virtue of their temperature; this radiation is called *thermal radiation* and depends in general both on the temperature of the body and on the nature of its surface.

A particularly interesting form of thermal radiation is that contained within a cavity in which the surrounding walls are at the same temperature *T*; this radiation is known as *cavity radiation*. Fig. 8.4 shows such a cavity. The thermal radiation will be absorbed and re-emitted by the walls of the cavity until an equilibrium state is reached with no further changes occurring in the nature of the radiation. Although at first sight it might

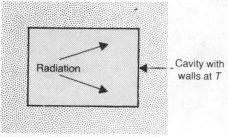

Fig. 8.4 Cavity or black body radiation.

seem an unlikely candidate, it is possible to treat the thermal radiation within a cavity as a thermodynamic *PVT* system. We may regard the radiation as having the temperature *T* of the walls and the volume *V* of the cavity. Further, we know from electromagnetic theory that the radiation exerts a pressure *P* on the walls of the box and also that it has associated with it an energy *U*

$$U = \int\limits_{V} \left(\frac{B^2}{2\mu_0} + \frac{\epsilon_0 E^2}{2} \right) \mathrm{d}V \qquad [8.30]$$

where *B* is the magnetic induction field of the radiation, *E* is the electric field and d*V* is an element of volume of the box. The integrand in equation [8.30] is the energy per unit volume and is called the *energy density, u*. Thus such cavity radiation has all the attributes of a *PVT* system.

For reasons that will be made clear in section 8.9, cavity radiation is also known as *black body radiation*.

8.5 The radiation pressure

There is a very simple connection between the radiation pressure P and u. This can be derived from classical electromagnetic theory, but it is simpler to use some elementary ideas from both quantum theory and the kinetic theory of gases to obtain this result. However, the physical origin of this pressure can be understood from our ideas in electromagnetism as follows.

Consider an electromagnetic wave incident on a boundary as in Fig. 8.5. The E_x field of the wave will induce a current density j_x in the wall. The B_y field then interacts with j_x to produce a force $\mathscr{F}_z$ on the wall in the z direction. This is the origin of the radiation pressure.

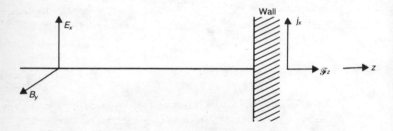

Fig. 8.5 The origin of radiation pressure according to the laws of electromagnetism.

Let us now obtain the relation between P and u. It is known from kinetic theory that the pressure in a gas is

$$P = 1/3 \; nm\overline{v^2} \qquad [8.31]$$

where n is the number density, m is the molecular mass and $\overline{v^2}$ is the mean square molecular velocity. As nm is the mass per unit volume or density ρ, we may rewrite equation [8.31] as

$$P = 1/3 \; \rho\overline{v^2} \qquad [8.32]$$

If we consider the radiation as a *photon gas* where the photons are all moving with the velocity c, equation [8.32] becomes

$$P = 1/3 \, \rho \, c^2 \qquad [8.33]$$

According to the Einstein mass-energy relation,

$$u = \rho \, c^2 \qquad [8.34]$$

Thus, equation [8.33] becomes

$$\boxed{P = 1/3 \, u} \qquad [8.35]$$

This is our required relation.

8.6 The spectral energy density, the absorptivity and the emissivity

Before we can proceed any further, we have to define the following.

1. *Spectral energy density* u_λ. Radiation in general does not consist of a single wavelength but a whole spectrum of wavelengths. Accordingly, we define the spectral energy density u_λ so that $u_\lambda d\lambda$ is the energy contained per unit volume between the wavelengths λ and $\lambda + d\lambda$. Clearly

$$u = \int_0^\infty u_\lambda \, d\lambda \qquad [8.36]$$

2. *Spectral absorptivity of a surface* α_λ. α_λ is defined as the fraction of energy incident on a surface that is absorbed at λ.

3. *Spectral emissivity of a surface* ϵ_λ. This is defined such that $\epsilon_\lambda d\lambda$ is the energy emitted per unit area per second by the surface between λ and $\lambda + d\lambda$.

8.7 The energy density for cavity radiation

In Fig. 8.6, we have a box, with the walls at T, in which there is cavity radiation. We know from common experience that the radiation emitted by the exterior of the box to the outside depends on both the temperature of the walls and the nature of the walls. For example, a red smooth box looks quite different from a blue rough box. However, there is something very *special*

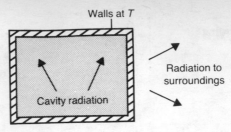

Fig. 8.6 The radiation within a cavity depends on the temperature of the walls only. The radiation from the outside of the walls depends on the nature of the walls as well as on their temperature.

about the radiation within the box. We make the following statement about this radiation, which we shall shortly prove at the end of this section:

> *Whatever the nature of the materials of the wall, the energy density for cavity radiation depends only on the temperature of the walls, while the spectral energy density depends only on the temperature and the wavelength.*

That is:

$$u = u(T) \qquad \text{[8.37]}$$

$$u_\lambda = u_\lambda(\lambda, T) \qquad \text{[8.38]}$$

The spectral energy curves for cavity radiation can be measured experimentally and it was in his attempt to understand them that Planck was led to his quantum theory in which he hypothesised that electromagnetic radiation is quantised in energy packets of size $h\nu$ where ν is the frequency of the radiation and h is a very small constant which we now know as Planck's constant. In fact Planck showed that the spectral energy density is given by

$$u_\lambda(\lambda, T) = \frac{\beta}{\lambda^5} \left(\frac{1}{e^{hc/\lambda k_B T} - 1} \right) \qquad \text{[8.39]}$$

where β is a constant. It can be seen that the Planck distribution law is consistent with our statement [8.38].

Plots of u_λ against λ for different temperatures are given in Fig. 8.7. The curves peak at a particular wavelength λ_{max} and this

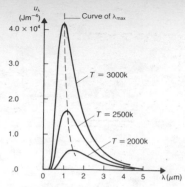

Fig. 8.7 The spectral energy density as a function of λ and T for cavity radiation. The curves follow the Planck distribution law.

wavelength moves to shorter wavelengths as the temperature is raised. It may be shown from the Planck distribution law, and also experimentally, that

$$\lambda_{max} T = \text{a constant} \qquad [8.40]$$

This result is known as Wien's law. It is interesting to note that the peak of the spectral distribution for the radiation from the sun, which is at a temperature of ~ 6000 K, is at $\lambda_{max} = 600$ nm. This is in the middle of the visible spectrum and so our eyes have evolved with the peak of the visual acuity at this wavelength.

The area under each spectral energy density curve gives the energy density for a particular temperature. It may be shown that integration of equation [8.39] gives

$$u = AT^4 \qquad [8.41]$$

where A is a constant. We shall derive this important result, using a simple thermodynamic argument, in section 8.8 where it will be related to the familiar Stefan radiation law which is well established experimentally. Before we do this, however, let us return to the statements [8.37] and [8.38] and prove their validity.

Consider the box shown in Fig. 8.8 which is composed of two halves, A and B, with the walls made of different materials but at the same temperature. Suppose that the energy density u in the two halves of the box is different with say $u_A > u_B$. If the two

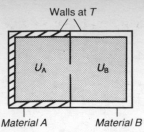

Walls at T

Fig. 8.8 If the energy density of the cavity radiation in A is different from that in B, then the second law may be shown to be violated. We conclude that the energy density of cavity radiation depends on the temperature only.

halves are separated by a partition with a hole in it, energy will be incident on both sides of the hole, with there being a net flux of energy from A to B as $u_A > u_B$. Thus B will heat and A will cool, with no external work being done on the system. This is a violation of the Clausius statement of the second law. A similar argument would hold if $u_B > u_A$. We conclude then that

$$u_A = u_B$$

In other words, the energy density within a cavity depends only on the temperature of the walls and not on the nature of the walls, which is the statement [8.37].

This argument may now be extended to the spectral energy density u_λ simply by covering the hole with a filter passing radiation only in the narrow band between λ and $\lambda + d\lambda$. u_λ^A must then equal u_λ^B by the argument just employed; from this, statement [8.38] follows.

8.8 The thermodynamic derivation of the Stefan Law

The energy equation for a PVT system is

$$\left(\frac{\partial U}{\partial V}\right)_T = T\left(\frac{\partial P}{\partial T}\right)_V - P \qquad [7.10]$$

Also, we have the following relations for cavity radiation:

$$P = 1/3\,u, \quad U = uV \quad \text{and} \quad u = u(T)$$

Substituting these equations in equation [7.10],

$$u = \frac{1}{3} T \frac{du}{dT} - \frac{1}{3} u$$

or $\quad \frac{4}{3} u = \frac{T}{3} \frac{du}{dT}$

or $\quad 4 \frac{dT}{T} = \frac{du}{u}$

Integrating,

$$u = A T^4 \qquad\qquad\qquad [8.42]$$

where A is a constant. This is the result we quoted in the previous section where we stated that it also follows from the Planck distribution law.

How then is the energy density u related to the energy radiated by a body? Let us concern ourselves for the moment with a very special type of body, *a black body*, which is defined as one which absorbs all the incident radiation so that $\alpha_\lambda^{black} = 1$ for all wavelengths.

Again we draw on our results for kinetic theory. We know that the number of molecules striking unit area per second is $1/4\ n\overline{v}$. Suppose the average energy carried by each photon is $\overline{E}$. Then our kinetic theory result implies that the radiation energy incident per second on unit area of a body placed within our cavity is $1/4\ nc\overline{E} = 1/4\ uc = 1/4\ cAT^4$ as $n\overline{E} = u = AT^4$. If the body is a black body, this is also the energy absorbed and, for equilibrium, must also be equal to the energy radiated. That is

The energy radiated per unit area per second by a black body is σT^4

where we have written σ for the constant $1/4\ Ac$. This is Stefan's law, with σ being known as the Stefan constant.

If the body is not black with a mean absorptivity α over all wavelengths with $\alpha < 1$, the above argument has to be modified to give the energy radiated as $\alpha\ \sigma T^4$ rather than σT^4.

8.9 The Kirchoff law

The Kirchoff law is often quoted as 'Good absorbers are good emitters.' A more precise formulation is that $\epsilon_\lambda/\alpha_\lambda$ *is a constant for all surfaces, at a given temperature and wavelength.*

Let us see how the Kirchoff law follows from what has previously been said. We refer again to Fig. 8.6. If we place a body (which is not necessary black) inside the cavity, the radiation will be preserved within the cavity if the energy absorbed by the body per second, between the wavelengths λ and $\lambda + d\lambda$, is equal to the energy radiated between those wavelengths. That is

$$\alpha_\lambda 1/4cu_\lambda \, d\lambda = \epsilon_\lambda \, d\lambda \qquad [8.43]$$

or

$$\frac{\epsilon_\lambda}{\alpha_\lambda} = \frac{1}{4} u_\lambda c \qquad [8.44]$$

so

$$\frac{\epsilon_\lambda}{\alpha_\lambda} = \frac{c}{4} u_\lambda(\lambda, T) \qquad [8.45]$$

using equation [8.38]. The right hand side of equation [8.45] is a universal function of λ and T and is independent of the nature of the body. We conclude that, *at a given wavelength and temperature,*

$$\epsilon_\lambda = (a \ constant) \ \alpha_\lambda \qquad [8.46]$$

where the constant is the same for all bodies. This is the Kirchoff law. Because good absorbers are good emitters or, conversely, bad absorbers are bad emitters, the runners in the London marathon are all provided with shiny foil capes at the end of the race to preserve their body heat.

Finally, there is one more important point. If the body is black, with $\alpha^{\text{black}} = 1$, equation [8.44] gives

$$\epsilon_\lambda^{\ \text{black}} = c/4 \, u_\lambda(\lambda, T)$$

So u_λ has exactly the same dependence on λ and T as does $\epsilon_\lambda^{\ \text{black}}$. *It is for this reason that we call cavity radiation, black-body radiation.*

RUBBER BANDS

8.10 The thermal expansion coefficient for rubber

For a stretched wire, we have seen in section 2.8 that the work performed on the system in stretching the wire through the infinitesimal distance dL is $\mathscr{F}\,dL$. Let us keep the pressure constant so that we would expect no change in the volume with no volume work being done. However, just stretching a wire can result in a change of volume and so we ought to include a $-P\,dV$ term in the first law for the volume work done against the surroundings. In practice, the change in volume of the wire is so small, even for a finite stretching of the wire, that we may ignore this $-P\,dV$ term in comparison with the $\mathscr{F}\,dL$ term. The first law is:

$$dU = dQ + \mathscr{F}\,dL$$

or $\quad dQ = dU - \mathscr{F}\,dL$ [8.47]

The thermodynamic treatment of a stretched wire then follows that for a PVT system, with P being replaced by $-\mathscr{F}$ and V by L. In particular, the four Maxwell equations for a stretched wire may instantly be produced from the modified mnemonic:

$$\begin{array}{ccc} & S & \\ - & & \\ -\mathscr{F} & & L \\ & T & \end{array}$$

The pair of minus signs is handled in exactly the same way as for the corresponding mnemonic in a magnetic system, with a constant $-\mathscr{F}$ appearing outside a partial differential being taken as a constant $+\mathscr{F}$. One such Maxwell equation is

$$\left(\frac{\partial S}{\partial L}\right)_T = -\left(\frac{\partial \mathscr{F}}{\partial T}\right)_L$$ [8.48]

This of course comes from the Helmholtz free energy.

Let us briefly discuss the special case of a stretched wire — a rubber band. If an X-ray photograph is taken of a rubber band, it is seen to be composed of very long chains of molecules. In the unstretched state, these chains are intertwined around each other so that there is a high degree of disorder. When the band is isothermally stretched, it is observed that the molecules untwine themselves with a kind of ordered crystalline arrangement being produced. Thus we expect the entropy to go down in such an isothermal stretching, with

$$\left(\frac{\partial S}{\partial L}\right)_T < 0 \qquad\qquad [8.49]$$

Using the Maxwell relation equation [8.48] in equation [8.49]

$$\left(\frac{\partial \mathscr{F}}{\partial T}\right)_L > 0 \qquad\qquad [8.50]$$

We can see what this implies physically by using the cyclical rule:

$$\left(\frac{\partial \mathscr{F}}{\partial T}\right)_L \left(\frac{\partial L}{\partial \mathscr{F}}\right)_T \left(\frac{\partial T}{\partial L}\right)_\mathscr{F} = -1$$

or $\qquad \left(\frac{\partial L}{\partial T}\right)_\mathscr{F} = -\left(\frac{\partial \mathscr{F}}{\partial T}\right)_L \left(\frac{\partial L}{\partial \mathscr{F}}\right)_T \qquad [8.51]$

The first partial differential on the right-hand side is positive, as we have just shown in equation [8.50], while the second is also always positive as all rubber bands increase their length upon an increase in tension. We conclude then that

$$\left(\frac{\partial L}{\partial T}\right)_\mathscr{F} < 0 \qquad\qquad [8.52]$$

which means that the coefficient of linear expansion for a rubber band *is negative*. You should test this for yourself. Hang a weight from the end of a rubber band and heat the band by playing a match flame up and down, being careful not to burn the rubber. The rubber band should shrink and the weight should rise!

A final amusing application of this idea is shown in Fig. 8.9. A bicycle wheel, with the heavy tyre removed, is mounted with the axle horizontal. The steel spokes have been replaced by rubber bands which are in tension so keeping the rim in place. When the right hand side of the wheel is warmed by illuminating it with an infrared health lamp, the wheel rotates anticlockwise. Can you explain this effect?

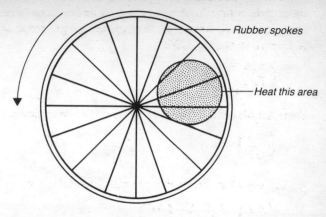

Fig. 8.9 When the rubber spokes on one side of the wheel are heated, the wheel rotates.

A REVERSIBLE ELECTROLYTIC CELL

8.11 The heat of reaction and the temperature coefficient of the *EMF*

As a final example of the application of thermodynamics to a non-PVT system, let us consider the important case of a reversible electrolytic cell, such as the Daniell cell, which drives a current through an external circuit as a result of the reaction

$$Zn + CuSO_4 \rightleftarrows Cu + ZnSO_4 \qquad [8.53]$$

If current is to be driven through an external circuit from the copper to the zinc electrode, this reaction proceeds in the direction of the upper arrow. We shall discuss the conditions under which such a reaction can be made reversible in practice at the end of this section. In general, during the reaction, there will be a change in volume of the cell, especially if gases are produced. The infinitesimal work term is then $-PdV + \mathcal{E}\,dZ$, using equation [2.12]. However, for many cells such as the Daniell cell, the change in volume is either zero or negligible, so that we may write the first law as

$$dU = dQ + \mathcal{E}\,dZ \qquad [8.54]$$

The thermodynamic relations for a *PVT* system can be transposed to those for a cell by replacing V with Z and P with $-\mathcal{E}$. In particular, the four Maxwell relations can be obtained from the modified mnemonic

$$
\begin{array}{ccc}
& S & \\
- & & \\
-\mathcal{E} & & Z \\
& T &
\end{array}
$$

There is a very useful relation that we shall now derive for the change in the internal energy during an isothermal process for such a cell and which we can relate to the heat of reaction. The combined first and second laws are

$$dU = T\,dS + \mathcal{E}\,dZ \qquad\qquad [8.55]$$

Writing $S = S(T, Z)$,

$$dS = \left(\frac{\partial S}{\partial T}\right)_Z dT + \left(\frac{\partial S}{\partial Z}\right)_T dZ$$

Substituting this in equation [8.55], and remembering that dT is zero as we are considering an isothermal process,

$$dU = \left\{ \mathcal{E} + T\left(\frac{\partial S}{\partial Z}\right)_T \right\} dZ = \left\{ \mathcal{E} - T\left(\frac{\partial \mathcal{E}}{\partial T}\right)_Z \right\} dZ$$

using the Maxwell relation $(\partial S/\partial Z)_T = -\,(\partial \mathcal{E}/\partial T)_Z$. For a saturated cell operated under reversible conditions, it is found that the EMF is dependent only on temperature and not on the charge stored. Hence

$$\Delta U = \left(\mathcal{E} - T\frac{d\mathcal{E}}{dT} \right)\Delta Z \qquad\qquad [8.56]$$

for a finite process. In particular, if one mole of each of the metal ions goes into solution, the charge driven by the cell through an external circuit is $\Delta Z = -n\,F_0$. Here n is the number of electrons transferred per ion, F_0 is the Faraday constant and the minus sign has been included as ΔZ is negative if the cell is discharging. Also, we know from equation [6.10] that, for a process taking place at constant pressure and where there is a negligible change of volume, $\Delta H = \Delta U$. We have then finally

$$\Delta H = -n\, F_0 \left(\mathscr{E} - T \frac{\mathrm{d}\mathscr{E}}{\mathrm{d}T} \right) \qquad [8.57]$$

for a reversible cell operated under isothermal and isobaric conditions. Equation [8.57] is important because it enables us to determine the thermodynamic properties of a cell *simply from a measurement of the EMF and its temperature coefficient*, both of which can be measured with great accuracy. For example, in the case of the Daniell cell, where $n = 2$, measurements at 273 K give $\mathscr{E} = 1.0934$ V and d $\mathscr{E}/\mathrm{d}T = -4.53 \times 10^{-4}$ VK^{-1}, so equation [8.57] gives $\Delta H = -2.35 \times 10^{5}$ J mol^{-1}. This should be compared with the result obtained in a calorimetric determination — a somewhat more difficult measurement. We know from equation [6.20] that, in a constant pressure process, ΔH is the heat of reaction. Calorimetric measurements of this heat of reaction give $\Delta H = -2.30 \times 10^{5}$ J mol^{-1}, which is in good agreement with the value obtained using equation [8.57]. The minus sign shows that heat is given out, so that the reaction is exothermic.

Finally, let us return to the question of reversibility. The cell reaction can be made reversible, very simply, by balancing $\mathscr{E}$ with an external EMF from a potentiometer. Under these conditions, the current flow can be made very small and any irreversible effects such as Joule heating can be minimised. In fact, the current can be reversed with a small change in the balancing EMF, provided by the potentiometer.

Chapter 9
Change of phase

9.1 Phase

We are all familiar with the fact that, if the temperature is raised, ice melts into water and then the water turns into steam. Ice, water and steam are the three *phases* of the substance we generally term water. We are also familiar with the fact that two of the phases can coexist in equilibrium with each other; a beaker at 0 °C and at atmospheric pressure can contain ice floating in water with the mass of the ice remaining constant. In fact, at one particular temperature and pressure all three phases may exist together. Strictly, a phase consists of a homogeneous region of the substance having definite boundaries; this is certainly so for our ice in water example.

9.2 *PVT* surfaces

Let us concern ourselves with a system, such as a simple fluid, where P, V and T are the appropriate state variables. We know that the equilibrium states of the system in a single phase are uniquely specified by two of these variables, and that P, V and T are connected by the equation of state. Specifying the pair P and V, for example, then fixes T. If these variables are plotted along three mutually perpendicular axes, the different equilibrium states of the system define a surface, called the PVT surface. The PVT surface for a typical substance is shown in the centre of Fig. 9.1.

The PVT surface appears very complicated, principally because it is three dimensional and so it is hard to visualise. It can best be

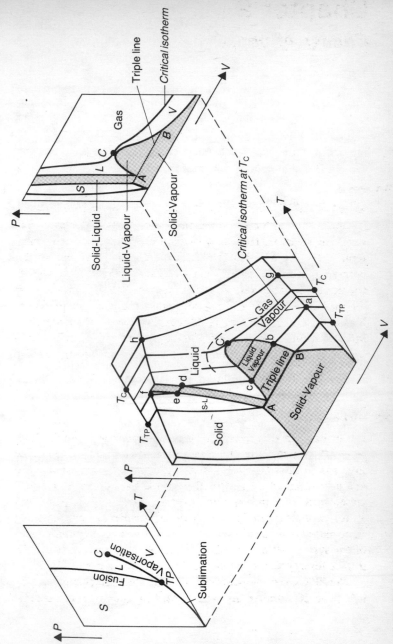

Fig. 9.1 A typical PVT surface together with its P–T and P–V projections.

understood as follows. Let us consider the isothermal path abcdef. Experimentally, we could take the system along this path by compressing it isothermally in a cylinder. At a, the system is in the single *vapour*-phase region. As the pressure is increased the volume decreases until, at b, we have condensation, with drops of *liquid* just beginning to appear; that is, the substance begins to separate into two distinct phases of quite different densities, although both are at the same temperature and pressure. As we go from b, where the substance is all vapour, to c, more and more liquid appears until, at c, the substance is all liquid; bc is thus in the two-phase liquid–vapour region. At b we say that we have a *saturated vapour*, while at c we have a *saturated liquid*.

c to d is in the single-phase liquid region. It requires a great increase in pressure to achieve a small change in volume as the compressibility of a liquid is generally small.

At d the substance begins to solidify, until at e it becomes all *solid*. de is in the two-phase solid–liquid region. The path e to f is in the single solid-phase region, where enormous pressures are generally required to effect a compression; thus the slope of ef is very large.

There is a *critical temperature* T_c above which an isothermal compression, such as the one we have been considering, produces no sharp liquid–vapour transition. Upon compression, the system becomes more and more dense, moving continuously from being a low-density fluid into a high-density fluid. The isotherm gh is an example of this. This effect would also be observed if the non-isothermal path, shown as the dotted line starting at a, is followed. Below T_c, as we have seen, it is possible for the system to exist in two separate phases, liquid and vapour, with quite different densities. At the *critical point*, C, the vapour and the liquid have become indistinguishable with the same density. It is customary to use the word *gas* above T_c and to use the word *vapour* below T_c. In other words, compressing a gas will not produce condensation.

The region marked S–V represents the two-phase solid–vapour region where no liquid is present. If we were to follow the system along an isotherm through this region in a similar way to the one just followed through the liquid–vapour region, we would pass from the vapour-only phase, through the two-phase solid–vapour phase and into the solid phase with no liquid ever having been encountered.

The L-V and S-V regions are separated by the isotherm at T_{TP}. At this temperature, and only at this temperature, all three phases may coexist, with the ratio of vapour to liquid to solid varying along BA. The line BA is called the *triple line* because of the coexistence of the three phases along it.

The projections of the *PVT* surface onto the *PV* and *PT* planes are indicated in Fig. 9.1 and also separately in Fig. 9.2. We shall continue our discussion of phase changes using these projections as they are easier to draw than the full three-dimensional surface. Because the portions of the isotherms in the two phase regions (such as bc and de in Fig. 9.1) are straight lines, they project onto the *PT* plane as points. In particular, the triple line projects into the *triple point* (TP); there is a unique triple point for each substance, except helium. The points from the projections of the other lines join up to form continuous curves which are the phase boundaries. We call the phase boundaries separating solid from vapour, solid from liquid and liquid from vapour the *sublimation*, *melting* and *vaporisation* curves respectively. We should realise

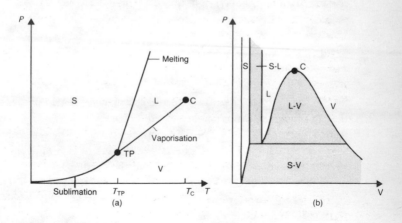

Fig. 9.2 P-T and P-V projections for a typical substances. S = solid; L = liquid; V = vapour.

that these curves are no more than the loci of the different sub-limation, melting and boiling points.

9.3 The equilibrium condition for two phases

Suppose we have a system consisting of two phases of a single substance, i.e. a single *component*. Such a system could be ice and water. Although a mixture of phenol and water consists of two phases, it has two different components (water and phenol) and so does not concern us here. If this system is maintained at a constant temperature and pressure, we know from section 6.5 that the condition for thermodynamic equilibrium is that the Gibbs function is a minimum.

Let us adopt the usual notation that extensive quantities per unit mass, or specific values, take lower case symbols. If the two phases are 1 and 2 with masses M_1 and M_2 so that the total mass $M = M_1 + M_2$,

$$G = M_1 g_1 + M_2 g_2 \tag{9.1}$$

At equilibrium,

$$dG = 0 = g_1 dM_1 + g_2 dM_2 \tag{9.2}$$

If the system is closed so that M is constant,

$$dM = dM_1 + dM_2 = 0 \tag{9.3}$$

Substituting equation [9.3] into equation [9.2],

$$\boxed{g_1 = g_2} \tag{9.4}$$

So the equilibrium condition for the coexistence of two phases is that the specific Gibbs functions are equal. At equilibrium, any amount M_1 $(M_1 \leqslant M)$ of phase 1 may coexist with the remaining amount $(M - M_1)$ of phase 2 because the value of G is unchanged as M_1 is altered. We can thus understand the existence of the two-phase regions of Fig. 9.1. The equality of g for the two phases at equilibrium is a very powerful result which now leads us directly to the Clausius–Clapeyron equation.

9.4 The Clausius–Clapeyron equation for first-order phase changes

A first-order phase change in a substance is characterised by a change in the specific volume between the two phases, accompanied by a latent heat. A solid melting into a liquid or a liquid boiling into a vapour are examples of first-order phase changes; first-order phase changes are in fact the familiar type of phase change. Consider the PT projection shown in Fig. 9.3.

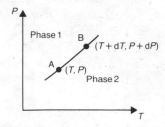

Fig. 9.3 A portion of a phase boundary.

At A, where the pressure and temperature are P and T,

$$g_1(T,P) = g_2(T,P) \qquad [9.5]$$

At the neighbouring state B, where the pressure and temperature are $P + dP$ and $T + dT$,

$$g_1(T + dT, P + dT) = g_2(T + dT, P + dT) \qquad [9.6]$$

Using Taylor's theorem, equation [9.6] becomes, to first order,

$$g_1(T,P) + \left(\frac{\partial g_1}{\partial T}\right)_P dT + \left(\frac{\partial g_1}{\partial P}\right)_T dP = g_2(T,P)$$
$$+ \left(\frac{\partial g_2}{\partial T}\right)_P dT + \left(\frac{\partial g_2}{\partial P}\right)_T dP$$

or $\quad \left\{\left(\frac{\partial g_1}{\partial T}\right)_P - \left(\frac{\partial g_2}{\partial T}\right)_P\right\} dT = \left\{\left(\frac{\partial g_2}{\partial P}\right)_T - \left(\frac{\partial g_1}{\partial P}\right)_T\right\} dP \quad [9.7]$

using equation [9.5]. As

$$v = \left(\frac{\partial g}{\partial P}\right)_T \quad \text{and} \quad s = -\left(\frac{\partial g}{\partial T}\right)_P, \qquad [6.40]$$

equation [9.7] becomes

$$(s_2 - s_1)\, dT = (v_2 - v_1)\, dP$$

or $\quad \dfrac{dP}{dT} = \dfrac{s_2 - s_1}{v_2 - v_1} = \dfrac{S_2 - S_1}{V_2 - V_1}$ [9.8]

It is important to realise that the quantities appearing in equation [9.8] have to refer to the *same mass* of the substance in the two phases. This could be a kilogram, a mole or even a molecule.

A phase change in a system occurring with a change in its entropy implies that there is a transfer of heat to or from the surroundings; this is the *latent heat L*. If a fixed mass of phase 1 changes into phase 2 at the temperature T, it follows that from equation [5.4] that $L = T(S_2 - S_1)$ or, referred to unit mass,

$$l = T(s_2 - s_1)$$ [9.9]

When $s_2 > s_1$, l is positive and heat has to be supplied to the system. Substituting equation [9.9] into equation [9.8],

$$\boxed{\dfrac{dP}{dT} = \dfrac{l}{T(v_2 - v_1)} = \dfrac{L}{T(V_2 - V_1)}}$$ [9.10]

This is the Clausius–Clapeyron equation for the *slope of the phase boundary*. Notice that, for this equation to make sense, there has to be a volume change between the two phases as well as a latent heat. These are precisely the conditions for a first-order phase change.

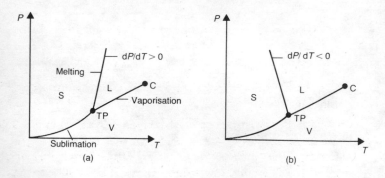

Fig. 9.4 (a) A P-T projection for a typical substance which expands on melting. (b) A P-T projection for a substance which contracts on melting.

In Fig. 9.4 (a) we show the PT projection for a substance which expands on melting from the solid to the liquid ($v_L > v_S$) and requiring latent heat for the phase change, so $s_L > s_S$. Then, from equation [9.10], dP/dT is positive, as shown. On the other hand, water contracts when ice melts into liquid and so the PT projection is as in Fig. 9.4(b) where dP/dT is negative. Water belongs to the small class of substances that behave in this way.

9.5 The melting point of ice and the boiling point of water

At $0\,^{\circ}\text{C}$ the specific volumes of ice and water are 1.09 cm^{3}g^{-1} and 1.00 cm^{3}g^{-1} respectively, while the latent heat of fusion is 335 J g^{-1}. Substituting these values into equation [9.10] we find that the slope of the fusion curve is $dP/dT = -134$ atmospheres K^{-1}. It is because of this negative slope that we can ice skate.

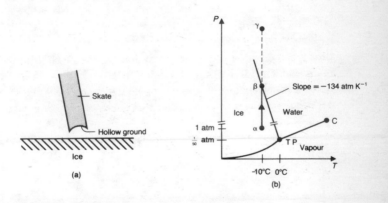

Fig. 9.5 The physics of skating. (a) Hollow-ground ice-skate. (b) *P–T* projection for pure water.

The bottom of an ice skate is hollow ground, as in Fig. 9.5(a), so an enormous pressure is built up under the sharp edge, of the order of 100 atmospheres or more. Suppose that the temperature is $-10\,^{\circ}\text{C}$. At one atmosphere the ice is in the state α on the PT projection of Fig. 9.5(b), and there is no water present. When the ice skater puts pressure on the ice, the state moves along the constant temperature line $\alpha\gamma$. However, as soon as the fusion curve is reached at β some ice melts so that the edge of the skate sinks in fractionally, with the load now being spread over a larger area, stabilising the pressure. The state thus remains fixed at β, with the liberated water acting as a lubricant. If the temperature is too cold, skating is impossible because the fusion curve cannot be reached with the increase in pressure, the starting point being too far to the left in Fig. 9.5(b).

It is often said that skiing is a pressure-melting effect too. This is nonsense as the pressure produced under the broad area of a ski is insufficient to produce pressure melting under normally encountered conditions. The snow is melted to produce the lubricating water simply by friction. Also, the bottom of skis are made slippery with wax.

Let us now consider the effect of pressure on the boiling point of water. The specific volumes of water and steam at 100 °C and 1 atmosphere are 1.043 cm^{3}g^{-1} and 1673 cm^{3}g^{-1} respectively, while the latent heat of vaporisation is 2257 J g^{-1}. If these values are substituted into equation [9.10] we find that dP/dT is $+ 1/28$ atmospheres K^{-1}. This means that, on the top of Mount Everest where the pressure is only 0.35 atmospheres, the boiling point of water is depressed by $0.65/(1/28)$ or 18 °C, so that the boiling point is only 82 °C.

In actual fact, the PVT surface of water is somewhat more complex than has been suggested, but this does not affect the validity of our previous discussion. The interested reader is referred elsewhere.

9.6 The equation of the vaporisation curve

We may very simply obtain an approximate equation for the vaporisation curve if we assume that the specific volume of the vapour v_V is very much larger than the specific volume v_L for the liquid and that the vapour obeys the equation of state for a perfect gas.

As $v_V \gg v_L$, and $Pv_V = RT$ taking our unit mass as one mole, equation [9.10] becomes

$$\frac{dP}{dT} \approx \frac{l}{Tv_V} = \frac{lP}{T^2 R} \qquad\qquad [9.11]$$

Integration then gives

$$\boxed{\ln P = -\frac{l}{RT} + \text{constant}} \qquad \text{(one mole)} \quad [9.12]$$

where we have assumed that l is constant over the region of the integration.

This equation tells us how the saturation vapour pressure varies with temperature. In fact, equation [9.12] is used in very low temperature thermometry, at less than 1 K, by relating the measurable pressure above a bath of liquid helium to the temperature.

Equation [9.12] may also be applied to the sublimation curve.

9.7 The variation of G in first-order transitions

There are some simple analytical arguments from which we can obtain information about the changes of entropy and volume in a first-order phase transition.

The change in S

Let us consider how g varies as we follow the constant P section XY in Fig. 9.6(a). We know that, as

$$\left(\frac{\partial g}{\partial T}\right)_P = -s \qquad\qquad [6.40]$$

the g versus T plot has a *negative slope* as s is always positive. Further,

$$\left(\frac{\partial^2 g}{\partial T^2}\right)_P = -\left(\frac{\partial s}{\partial T}\right)_P = -\frac{c_P}{T} < 0 \qquad\qquad [9.13]$$

as c_P is positive, so such a plot always bends down *towards* the T axis. We know that g is a smoothly varying function of T and P

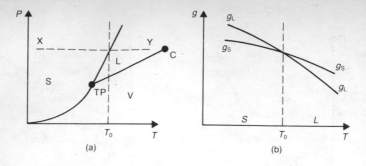

Fig. 9.6 (a) An isobaric section XY across a P–T projection. (b) We argue in the text that the Gibbs function varies as shown here.

which can take different values for all T and P. Fig. 9.6(b) shows the Gibbs functions for the solid and liquid phases. At T_0 we know that they are equal so the curves cross at that point. Also, for $T < T_0$, the solid phase is the stable phase. This means that its g curve must be the lower one in this region to minimise the Gibbs function for the system. The higher g curve for the liquid phase represents the unstable phase. g for the constant P section varies then as in Fig. 9.7(a), with a discontinuity in the slope at T_0. This implies a discontinuity in s as shown in Fig. 9.7(b), with the *higher temperature phase having the greater s*.

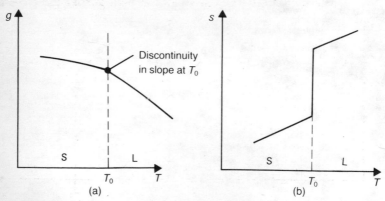

Fig. 9.7 (a) The variation of g with T along the section XY of Fig. 9.6. (b) The corresponding behaviour of the entropy. Notice that the high-temperature phase has the higher entropy.

The change in V

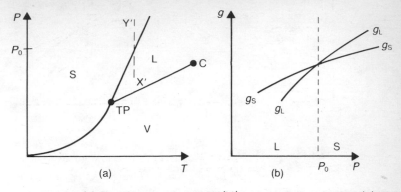

Fig. 9.8 (a) An isothermal section X'Y' across a P-T projection. (b) We argue in the text that the Gibbs function varies as shown here.

If we were instead to consider the constant T section X'Y' of Fig. 9.8(a), a similar argument gives the plots of g against T for the solid and liquid phases, as shown in Fig. 9.8(b), because:

$$\left(\frac{\partial g}{\partial P}\right)_T = v > 0 \quad \text{and} \quad \left(\frac{\partial^2 g}{\partial P^2}\right)_T = \left(\frac{\partial v}{\partial P}\right)_T < 0,$$

as all known substances suffer a decrease in volume upon an increase of pressure. Thus the Gibbs function varies as in Fig. 9.9(a)

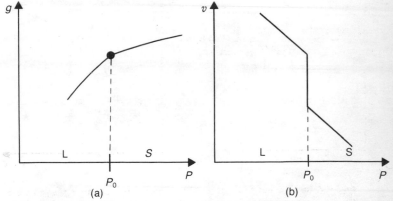

Fig. 9.9 (a) The variation of g with P along the section X'Y' of Fig. 9.7. (b) The corresponding behaviour of the specific volume. Notice that the high-pressure phase has the smaller specific volume, as expected.

with a discontinuity in the slope at $P = P_0$ — in other words in v. This implies that v varies as in Fig. 9.9(b), with the *high-pressure phase having the smaller specific volume*, a result consistent with experience.

9.8 Second-order phase changes

We have seen, in a first-order phase change: there is a change in the specific volume; there is a latent heat, which means that there is a change in the specific entropy; and there is no change in the specific Gibbs function. That is

$$g_1 = g_2, \quad v_1 = v_2 \quad \text{and} \quad s_1 = s_2$$

As $\quad s = -\left(\dfrac{\partial g}{\partial T}\right)_P \quad$ and $\quad v = \left(\dfrac{\partial g}{\partial P}\right)_T \qquad$ [6.40]

we can very elegantly describe a first-order phase change as one in which g is continuous, but the first-order derivatives of g with respect to the natural variables of P and T are discontinuous.

Second-order phase changes can be defined by extending this classification. In a second-order phase change, g is continuous; the first-order derivatives of g with respect to T and P are now continuous so that there is *no change in the specific volume and there is no latent heat*; but there are discontinuities in the second-order derivatives. That is

$$g, \quad s \quad \text{and} \quad v \quad \text{are continuous}$$

but $\quad \left(\dfrac{\partial s}{\partial T}\right)_P, \quad \left(\dfrac{\partial s}{\partial P}\right)_T, \quad \left(\dfrac{\partial v}{\partial T}\right)_P \quad \text{and} \quad \left(\dfrac{\partial v}{\partial P}\right)_T$

are discontinuous. As

$$c_P = T\left(\dfrac{\partial s}{\partial T}\right)_P, \quad \beta = \dfrac{1}{v}\left(\dfrac{\partial v}{\partial T}\right)_P \quad \text{and}$$

$$\kappa = -\dfrac{1}{v}\left(\dfrac{\partial v}{\partial P}\right)_T,$$

this means that c_P, β and κ are discontinuous. The second and third partial differentials are the same, by the Maxwell relation, equation [6.41], apart from a difference in sign which is of no importance as we are talking only of continuity.

This classification of phase changes can be further extended to third, and even higher, order phase changes and is known as the Ehrenfest classification.

9.9 Examples of phase changes of different orders

There are many examples of first- and second-order phase changes, or transitions, in physics, metallurgy and chemistry, and we must unfortunately restrict ourselves to listing only a few in each category.

First-order phase changes

1. A solid melting into a liquid, a liquid boiling into a gas, and a solid subliming into a gas are the most familiar examples of a first-order phase change.
2. Below a certain characteristic temperature, the critical temperature T_c, certain metals become superconductors. This means that, whereas they behave as normal metals at temperatures above T_c, exhibiting electrical resistance, below T_c they have no electrical resistance at all! A current circulating in a ring of lead would go on circulating for ever provided the lead was kept cold at a temperature less than 7.2 K. The values for T_c for a few superconductors are given in Table 9.1.

Table 9.1 *Values of* T_c *for a few superconductors*

Superconductor	Critical temperature T_c (K)	Critical field $B_{0C}(0)$ (tesla)
Niobium	9.2	0.26
Lead	7.2	0.081
Mercury	4.2	0.042
Tin	3.7	0.031

The superconducting phase can be destroyed by raising the temperature above T_c in the absence of a magnetic field. At any $T < T_c$ the superconducting phase may also be destroyed by an applied magnetic induction above a certain critical value of $B_{oc}(T)$; this critical field increases as the temperature is reduced, reaching the value $B_{oc}(0)$ as $T \to 0$. This is illustrated for a typical superconductor in the phase diagram shown in Fig. 9.10, where the phase boundary separates the normal from the superconducting phase. The third column of Table 9.1 gives the values of $B_{oc}(0)$ for different superconductors.

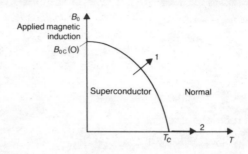

Fig. 9.10 A phase diagram for a typical superconductor. The transition 1, occurring in an applied field, is a first-order phase transition, while that denoted by 2, occurring in the absence of a field, is an example of a second-order phase transition.

The transition from a superconductor to a normal conductor is a first-order phase change, *provided the transition takes place in an applied magnetic field.* We have indicated such a transition by the arrow 1 in Fig. 9.10.

Second-order phase changes

1. Magnetic materials fall broadly into three main classes; the *diamagnetic class*, where the atoms of the material have no permanent magnetic moment but one is induced on the application of a magnetic field; the *paramagnetic class*, where the atoms have a permanent magnetic moment but it requires the application of an external magnetic field to destroy the random orientation of the individual moments and to give the material a net magnetic moment; and the *ferromagnetic* class, where the atoms have a net magnetic moment and these couple

together to give the material a net moment even in the absence of an external magnetic field. As the temperature of a ferromagnet is raised, it becomes a paramagnet at the *Curie temperature*. This transition is a second-order phase change.

2. The transition of a superconductor to a normal conductor is a second-order phase change, provided the transition does *not take place in an applied magnetic field*. Then, of course, the transition is of first order. The second-order phase change is indicated by the arrow 2 in Fig. 9.10.

3. Perhaps the most dramatic example of a second-order phase change occurs in liquid helium as the temperature is lowered through 2.2 K, where it changes from being a normal liquid to becoming a superfluid liquid with quite extraordinary properties. As the name implies, the superfluid phase is marked by a complete absence of any internal friction, that is viscosity. A rotating mass of superfluid helium would go on rotating for ever, in exact analogy to the persistence of the current in a ring of superconductor, provided that the velocity is below a certain critical limit.

   Let us first be clear about the notation. Helium can exist as two isotopes: the common isotope of mass number 4, ^{4}He, and commonly called helium four; and the rare lighter isotope of mass number 3, ^{3}He, commonly called helium three. We are referring *here* to the two phases of helium four, but

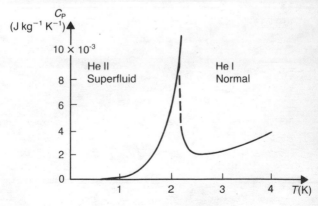

Fig. 9.11 The λ anomaly in the heat capacity of ^{4}He.

unfortunately the two phases are confusingly called Helium I for the normal phase and Helium II for the superfluid phase, a numbering system which has nothing to do with the mass number.

A plot of the heat capacity of ^{4}He against temperature is shown in Fig. 9.11. The discontinuity in the heat capacity at 2.2 K is clear. Because of the resemblance of this curve to the Greek letter lambda, the transition temperature is called the λ *point*.

9.10 The Ehrenfest equations for second-order phase changes

There exist two simple relations for the slope dP/dT of the phase boundary in a second-order phase change. We shall derive these two relations in an exactly analogous manner to that used in deriving the Clausius–Clapeyron equation for the slope of the phase boundary in a first-order phase change. The Clausius–Clapeyron equation does not give the slope of the phase boundary for a second-order phase transition because both l and Δv vanish, with equation [9.10] giving an indeterminate answer for dP/dT.

Let us return to Fig. 9.3, where there is now a second-order phase change between the two phases and the solid line is the phase boundary for this transition. As before, we consider the two neighbouring points A and B on the phase boundary at (T, P) and $(T + dT, P + dP)$. We recall that, in a second-order phase change, there is no change in either s or v in going from one phase to another.

Let us consider s first. We have:

At A, $\quad s_1(T, P) = s_2(T, P)$ [9.14]

At B, $\quad s_1(T + dT, P + dP) = s_2(T + dT, P + dP)$ [9.15]

Using Taylor's theorem, equation [9.15] becomes, to first order,

$$s_1(T, P) + \left(\frac{\partial s_1}{\partial T}\right)_P dT + \left(\frac{\partial s_1}{\partial P}\right)_T dP = s_2(T, P)$$

$$+ \left(\frac{\partial s_2}{\partial T}\right)_P dT + \left(\frac{\partial s_2}{\partial P}\right)_T dP$$

or $\quad \left(\dfrac{\partial s_1}{\partial T}\right)_P dT + \left(\dfrac{\partial s_1}{\partial P}\right)_T dP \ = \ \left(\dfrac{\partial s_2}{\partial T}\right)_P dT + \left(\dfrac{\partial s_2}{\partial P}\right)_T dP$

$$[9.16]$$

using equation [9.14]. Multiplying equation [9.16] all through by T and remembering that

$$c_P = T \left(\frac{\partial s}{\partial T}\right)_P, \quad \beta = \frac{1}{V}\left(\frac{\partial V}{\partial T}\right)_P = \frac{1}{v}\left(\frac{\partial v}{\partial T}\right)_P,$$

$$v_1 = v_2 \ ,$$

and $\quad \left(\dfrac{\partial s}{\partial P}\right)_T = -\left(\dfrac{\partial v}{\partial T}\right)_P,$ $\qquad\qquad$ [6.41]

we have

$$c_{P1} \ dT - Tv\beta_1 dP = c_{P2} \ dT - Tv\beta_2 dP \qquad\qquad [9.17]$$

where the subscripts on c and β refer to the two phases. Solving for dP/dT in equation [9.17], we have finally

$$\boxed{\dfrac{dP}{dT} = \dfrac{c_{P1} - c_{P2}}{Tv(\beta_1 - \beta_2)} = \dfrac{C_{P1} - C_{P2}}{TV(\beta_1 - \beta_2)}} \qquad\qquad [9.18]$$

This is the first Ehrenfest equation. As for the Clausius–Clapeyron equation, the extensive quantities (here C_P and V) have to refer to the same mass of the substance in each phase.

Let us now consider the continuity of v.

At A, $\quad v_1(T,P) = v_2(T,P)$ $\qquad\qquad\qquad$ [9.19]

At B, $\quad v_1(T + dT, P + dP) = v_2(T + dT, P + dP)$ $\qquad$ [9.20]

Using Taylor's theorem, equation [9.20] becomes, to first order,

$$v_1(T,P) + \left(\frac{\partial v_1}{\partial T}\right)_P dT + \left(\frac{\partial v_1}{\partial P}\right)_T dP = v_2(T,P)$$

$$+ \left(\frac{\partial v_2}{\partial T}\right)_P dT + \left(\frac{\partial v_2}{\partial P}\right)_T dP$$

or $\quad \left(\dfrac{\partial v_1}{\partial T}\right)_P \mathrm{d}T + \left(\dfrac{\partial v_1}{\partial P}\right)_T \mathrm{d}P = \left(\dfrac{\partial v_2}{\partial T}\right)_P \mathrm{d}T + \left(\dfrac{\partial v_2}{\partial P}\right)_T \mathrm{d}P$

[9.21]

using equation [9.19]. Remembering that

$$\beta = \frac{1}{V}\left(\frac{\partial V}{\partial T}\right)_P = \frac{1}{v}\left(\frac{\partial v}{\partial T}\right)_P \quad \text{and} \quad \kappa = -\frac{1}{V}\left(\frac{\partial V}{\partial P}\right)_T$$

$$= -\frac{1}{v}\left(\frac{\partial v}{\partial P}\right)_T \quad,$$

equation [9.21] becomes

$$\beta_1 v_1 \mathrm{d}T - \kappa_1 v_1 \mathrm{d}P = \beta_2 v_2 \mathrm{d}T - \kappa_2 v_2 \mathrm{d}P \qquad [9.22]$$

Collecting terms, and remembering that $v_1 = v_2$,

$$\boxed{\dfrac{\mathrm{d}P}{\mathrm{d}T} = \dfrac{\beta_2 - \beta_1}{\kappa_2 - \kappa_1}} \qquad [9.23]$$

This is the second Ehrenfest equation.

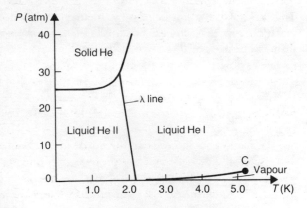

Fig. 9.12 The phase diagram for ^{4}He. The normal liquid phase is called He I while the superfluid phase is called He II.

A very nice example of the application of the first Ehrenfest equation is in determining the slope of the phase boundary between He I and He II. Fig. 9.12 is the phase diagram for ^{4}He where the λ line is the phase boundary for the second-order phase change between the two liquid phases. Notice that the solid phase is only ever reached upon the application of considerably greater pressures than atmospheric pressure. Measurements give the values of c_P and β for the two phases. Equation [9.18] then gives $dP/dT = -78$ atmospheres K^{-1}, which is in agreement with value of the slope determined in other ways.

9.11 Superconductivity and superfluidity

We shall conclude this chapter with a very brief account of the theoretical explanation of superconductivity and superfluidity, attempting to draw the parallels between them.

According to quantum mechanics, particles in a restricted geometry exist in discrete quantum states with different energies, as in Fig. 6.3. In the condensed state, under conditions of high particle number-density and low temperatures, quantum effects become manifestly important, particularly so for electrons in a metal and the light atoms of liquid ^{4}He and liquid ^{3}He.

It is known that, because of the indistinguishability of atomic particles, such particles can be either *bosons* or *fermions*. There is no restriction on the number of bosons that can occupy a single quantum state; however, only one fermion can occupy each quantum state. Atomic particles have the property of *spin* which is a measure of their intrinsic angular momentum. Particles with half-integral spin behave as fermions while those with integral spin behave as bosons. Electrons, which have half-integral spin, are thus fermions. In contrast, the nuclei of ^{4}He, which have zero and therefore integral spin (two protons with opposed spins and two neutrons also with opposed spins), are bosons. However, the nuclei of ^{3}He, with one less neutron than ^{4}He, have half-integral spin and so form a fermion system.

Let us first see how the superfluidity in ^{4}He occurs. As the temperature is lowered, all the ^{4}He atoms pack into the lower-energy quantum states until, at a sufficiently low temperature, they are all in the same ground state. Being bosons, they can do this. Having all the ^{4}He atoms in the same quantum state results in superfluidity.

How then do we get superconductivity? The electrons are precluded from all packing into the same quantum state because they are fermions. It turns out however that the most energetic and important electrons can move as *pairs* through the metal lattice, with their spins opposed so that these pairs behave as bosons with zero spin. The pair separation is ~100 atomic diameters. These so-called *Cooper pairs* can all occupy the same lowest energy quantum state at low temperatures and this gives rise to superconductivity.

The question now arises as to whether a similar pairing of the nuclei of ^{3}He can occur so that these ^{3}He pairs behave as a boson system, giving rise to a superfluid phase. The theory here preceded the experimental evidence. Theory shows that such pairing *can* occur, except that the half-integral spins add to give a spin of one for the pair and thus a boson system. The superfluid phase in ^{3}He was found at 3 a.m. on Thanksgiving Day at Cornell University in 1971, but not until the temperature had been reduced to the incredibly low value of 2.6×10^{-3} K. In fact, a second superfluid

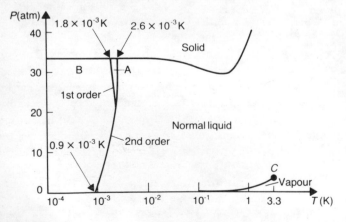

Fig. 9.13 The phase diagram for ^{3}He.

phase was also found at the even lower temperature of 1.8×10^{-3} K. The Cornell experimenters were actually following the solid–liquid transition as a function of temperature when they made their discovery.

The phase diagram for ^{3}He is given in Fig. 9.13. The boundary between the normal liquid and the two superfluid phases A and B denotes a second-order transition while the boundary between the two superfluid phases denotes a first-order transition. Below 10^{-3} K the vapour pressure is so small that it cannot be shown in Fig. 9.13.

Chapter 10
Open systems and the chemical potential

10.1 The chemical potential

There are many physical systems in which the quantity of matter is not fixed; in this short chapter we shall see how such systems are treated in thermodynamics by giving a very brief introduction to the concept of the chemical potential. As an example of such a system we could imagine a block of ice floating in water; as the ice melts its mass decreases because there is a transfer of H_2O molecules across the boundary dividing the ice from the liquid water. Alternatively we may consider a chamber containing a small hole through which we allow gas to enter from the surroundings; the gas in the chamber is then a system of variable mass.

For such a variable mass system, we must modify our central equation of thermodynamics,

$$dU = T\,dS - P\,dV \qquad [5.10]$$

to allow for the extra energy brought into the system by the additional particles that is locked up inside them. This extra energy is clearly of importance if it can be released to the rest of the system, and this could be the case if the particles were involved in, say, a chemical reaction of some form. Suppose for a moment that the system consists of only one type of particle and dN extra particles are added. Then can we rewrite equation [5.10] as

$$dU = TdS - P\,dV + \mu dN \qquad [10.1]$$

where μ is the so-called chemical potential which is defined as the increase in the internal energy per particle added under conditions of constant S and V:

$$\mu = \left(\frac{\partial U}{\partial N}\right)_{S,V} \qquad [10.2]$$

If there is more than one type of particle, equation [10.1] has to be modified (see question 1, Chapter 10, Appendix 4) to

$$dU = T\,dS - P\,dV + \sum_i \mu_i dN_i \qquad [10.3]$$

where the chemical potential for the ith type of particle is $\mu_i = (\partial U/\partial N_i)_{S,V,N_k}$ and the symbol N_k means that all the other N's except N_i are held constant. We shall assume for the moment that there is only one type of particle present, and shall extend our argument to the more general case of the presence of different types of particle when this is appropriate. It has to be remarked that, even if a particular type of particle is *initially* absent from the system, that does not mean that the corresponding μ is zero because it is a measure of the effect on U brought about by the *addition* of that type of particle.

Although our definition of μ has a clear physical interpretation, equation [10.2] is often inconvenient because of the requirement of constant entropy. Another definition of μ can be given in terms of F. As

$$F = U - TS \qquad [6.21]$$

and $\quad dF = dU - T\,dS - S\,dT$

it follows using equation [10.1] that

$$dF = -P\,dV - S\,dT + \mu\,dN \qquad [10.4]$$

so $\quad \boxed{\mu = \left(\frac{\partial F}{\partial N}\right)_{V,T}} \qquad [10.5]$

That is, μ is the increase of the Helmholtz free energy upon the addition of one particle under conditions of constant T and V.

This form of μ is important because it may be linked to the canonical distribution of statistical mechanics as

$$\mu = -k_B T \left(\frac{\partial}{\partial N}(N \ln z)\right)_{V,T} \qquad [10.6]$$

using equation [6.36]. Similarly, as

$$G = U + PV - TS \qquad [6.37]$$

it follows, using equation [10.1] again, that

$$dG = V\,dP - S\,dT + \mu dN \qquad [10.7]$$

so

$$\boxed{\mu = \left(\frac{\partial G}{\partial N}\right)_{T,P}} \qquad [10.8]$$

which is yet another form and, as we shall see, is the most useful. In other words, μ is the increase of the Gibbs free energy upon the addition of one particle under conditions of constant T and P. Notice that, in each case, the definition of μ is the partial differential of the thermodynamic potential with respect to particle number, with the appropriate natural variables being held constant.

Let us now see why equation [10.8], which gives the definition of μ in terms of G, is of particular importance. We know that G is an *extensive* quantity, that is it must be proportional to the particle number. Hence we may write

$$G(T,P,N) = N\,\phi(T,P) \qquad [10.9]$$

where $\phi(T, P)$ depends on the particular system being considered. Differentiating this with respect to N, keeping P and T constant,

$$\mu = \phi(T,P) \qquad [10.10]$$

which is *independent of N*. Also, it follows from equation [10.9] that

$$\mu = (G/N) \qquad \text{(one type of particle present)} \qquad [10.11]$$

Thus μ is no more than the Gibbs free energy per particle, provided only one type of particle is present. This is a more useful statement than the one given in our original definition, equation [10.8], which refers to the incremental increase in the Gibbs free energy per particle under conditions of constant T and P. In general, when there are a number of different types of particle present (different *particle types*), this incremental increase will depend on the existing particle populations and equation [10.11] has to be modified, as we shall shortly see, to equation [10.16].

On the other hand, we cannot obtain similar simple results for μ in terms of U and F. μ is *not* U/N or F/N because the natural variables for U and F are not *both* extensive as they are for G. For example, consider U. As U is extensive, we must write

$$U = U(S, V, N) = N\phi'\left(\frac{S}{N}, \frac{V}{N}\right) \tag{10.12}$$

where the new function ϕ' is of the entropy and volume per particle. Differentiating with respect to N at constant S and V gives μ again:

$$\mu = \phi' + N\left(\frac{\partial\phi'}{\partial N}\right)_{S,V} = \frac{U}{N} + N\left(\frac{\partial\phi'}{\partial N}\right)_{S,V} \tag{10.13}$$

So μ is not simply U/N but involves extra terms dependent on the particle number.

Another way of seeing the special relation between μ and G is to integrate equation [10.3]. Although this can be done rather elegantly using Euler's theorem for homogeneous functions, we shall use a less formal method. Let us scale up the system by a factor $(1 + \alpha)$, where α is a small number $\ll 1$. Then, all the extensive quantities, U, S, V, N_1, N_2 . . ., would increase in the same proportion with

$$\mathrm{d}U/U = \mathrm{d}S/S = \mathrm{d}V/V = \mathrm{d}N_i/N_i = \alpha \tag{10.14}$$

for all i. Substituting these relations for $\mathrm{d}U$, $\mathrm{d}S$ and $\mathrm{d}N_i$ in equation [10.3] and cancelling the common term α,

$$U = TS - PV + \sum_i \mu_i N_i \tag{10.15}$$

or

$$\boxed{G = \sum_i \mu_i N_i} \tag{10.16}$$

Equation [10.16] is a generalisation of our earlier result, equation [10.11]. It reduces to this if all the N_i except one are put equal to zero.

10.2 The equilibrium condition for two systems under particle exchange — phase equilibrium

Let us approach the concept of the chemical potential by enquiring what are the conditions for equilibrium between two systems which are allowed to exchange particles. We shall focus our discussion on change of phase by letting our two systems be two phases of the same substance. Then, each system will contain the same particles, all of which are identical, i.e. there is a single *particle type* throughout both systems. As an example of this, we could consider ice melting in water; then there is an exchange of particles (the water molecules) between the two systems. What are the general conditions of two phases to be in equilibrium against this particle exchange?

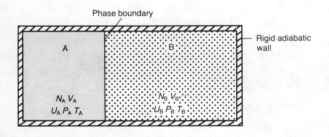

Fig. 10.1 A chamber containing two phases of the same substance. There can be particle exchange and heat flow across the boundary. Additionally, the boundary can move freely.

In Fig. 10.1 we have two phases, A and B, of the same substance occupying a chamber with rigid adiabatic walls. The two phases have the separate volumes V_A and V_B and are separated by a phase boundary. There can be heat and particle flow across this boundary. We shall allow this boundary to move, as of course it will do, for example, as ice melts into water. Let there be N_A particles in the phase A which has volume V_A, internal energy U_A and pressure P_A, and let there be a similar notation for the phase B. These quantities are subject to the conservation conditions:

$$N_A + N_B = N \quad \text{(the total number of particles)} \quad [10.17]$$

$$V_A + V_B = V \quad \text{(the total volume)} \quad [10.18]$$

$$U_A + U_B = U \quad \text{(the total internal energy)} \quad [10.19]$$

By the first law, U is fixed as the chamber has rigid adiabatic walls. Now, at equilibrium, we know that the entropy for the thermally isolated *combined* system of the two phases is a maximum:

$$S = S(U_A, V_A, N_A, U_B, V_B, N_B)$$
$$= S_A(U_A, V_A, N_A) + S_B(U_B, V_B, N_B) = \text{a maximum}$$

In an infinitesimal departure from this equilibrium state in which all the quantities U_i, V_i and N_i change by infinitesimal amounts from their equilibrium values,

$$dS = dS_A + dS_B = 0 \quad [10.20]$$

Applying equation [10.1] to the two phases A and B,

$$dS_A = \frac{1}{T_A}(dU_A + P_A dV_A - \mu_A dN_A) \quad [10.21]$$

$$dS_B = \frac{1}{T_B}(dU_B + P_B dV_B - \mu_B dN_B) \quad [10.22]$$

Substituting these values into equation [10.20] and using our conservation conditions where V, U and N are fixed so that $dV_A = -dV_B$, $dU_A = -dU_B$ and $dN_A = -dN_B$, we have

$$\left(\frac{1}{T_A} - \frac{1}{T_B}\right) dU_A + \left(\frac{P_A}{T_A} - \frac{P_B}{T_B}\right) dV_A$$
$$- \left(\frac{\mu_A}{T_A} - \frac{\mu_B}{T_B}\right) dN_A = 0 \quad [10.23]$$

This must be true for any dU_A, dV_A and dN_A and so their co-efficients in equation [10.23] must all be zero. Thus:

$T_A = T_B$,   which is the condition for thermal equilib-rium;

$P_A = P_B$,   which is the condition for mechanical equi-librium;

and $\mu_A = \mu_B$,   which is the condition for equilibrium against particle exchange.

We conclude finally:

> If we have two phases or systems in thermal and mechanical equilibrium then they will also be in equilibrium against particle flow and so in complete equilibrium if the chemical potentials are equal.

Just as heat flows from regions of high to low temperatures, particles will flow from regions of high to low chemical potential. We may see this as follows. Suppose we have a state very close to the final equilibrium state, differing only in that there is a small *positive excess* δN_A of particles in volume A over the equilibrium value N_A. There will be a corresponding deficit of $-\delta N_A$ in volume B. The entropy change in *returning* to the equilibrium state of maximum entropy must be slightly *positive* and is given by the sum of the last terms on the right of equations [10.21] and [10.22]. We must remember that the change in the population of A in *returning* to the equilibrium state is $-\delta N_A$ and that of B is $+\delta N_A$; there is also no change in the temperature. Hence

$$\delta S = -\mu_A \left(-\frac{\delta N_A}{T}\right) - \mu_B \left(+\frac{\delta N_B}{T}\right) > 0 \qquad [10.24]$$

or $$(\mu_A - \mu_B)\,\delta N_A > 0 \qquad [10.25]$$

As δN_A is positive, equation [10.25] shows that $\mu_A > \mu_B$, which means that the *particle flow is from the region of high chemical potential to the region of low chemical potential*.

We have in fact already met the equilibrium condition of the equality of the chemical potentials in the previous chapter, although in a disguised form. There we saw that the condition for the equilibrium of two phases at a fixed T and P is that the Gibbs functions per unit mass are equal. Now this means that the Gibbs functions per molecule are also equal for the two phases as they are composed of identical molecules. Further, as each phase is composed of only a single type of molecule or particle type, it follows from equation [10.11] that the Gibbs function per molecule is actually the chemical potential for each phase. Thus, for change of phase, the equality of the specific Gibbs functions is no more than our equality of the chemical potentials for equilibrium against particle flow. Of course, this should not surprise us as both results were derived from the same idea – the principle of increasing entropy.

Finally, we remark that we have chosen to discuss here, because it is so important, change of phase in which the boundary between our two systems is allowed to move. The simpler problem of the equilibrium conditions for two systems separated by a permeable diathermal wall which is *fixed* is left as question 2, Chapter 10 in Appendix 4. It will be seen there that the condition for equilibrium against particle flow is again the equality of the chemical potentials.

10.3 Three applications of the concept of the chemical potential

The flow of matter from one system to another occurs so frequently in nature that the concept of the chemical potential finds wide applicability. We have chosen to illustrate its use by considering three applications from the fields of biology, solid state physics and chemistry.

Osmotic pressure

We now extend our discussion to include *different particle types* in the chambers. Suppose we have a chamber, as in Fig. 10.2, separated into two regions A and B by a *rigid* diathermal wall that can sustain a pressure difference. Let A contain initially particles of type 1 only and B initially particles of type 2 only; this is shown in (a) of the figure. To be specific, we shall consider the two particles as forming gases. Let the wall be permeable to

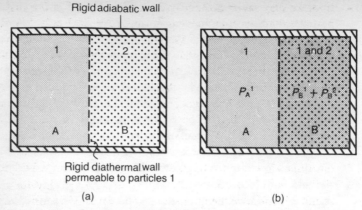

1

2

$P_A{}^1$

1 and 2

$P_B{}^1 + P_B{}^2$

A

B

A

B

Rigid diathermal wall
permeable to particles 1

(a)   (b)

Fig. 10.2 The semipermeable membrane separating the two regions A and B of the chamber is permeable to particles 1 only. This sets up an osmotic pressure across the membrane.

gas 1 only. Such a wall is called a *semipermeable membrane* and these occur in nature, particularly in biological systems. Such a membrane functions because it contains small holes which let through the smaller particles 1, say, but not the larger particles 2. Although biological membranes are not strictly rigid, after an initial deformation, they are able to sustain a pressure difference and may be regarded as rigid in that sense for the purpose of our discussion. Why, then, is a pressure difference set up across a membrane such as the one shown in Fig. 10.2? We may see the origin of this *osmotic pressure* as follows:

Particles of gas 1 will pass from A though the membrane into B until finally there is an equilibrium mixture of both types in B but only type 1 in A; this is indicated in Fig. 10.2(b). We shall use a superscript and subscript notation so that, for example, $P_A{}^1$ means the pressure due to type 1 particles in region A. You are asked to show in question 4, Chapter 10, Appendix 4 that, at equilibrium, $\mu_A{}^1 = \mu_B{}^1$, which is not a surprising result as it is clearly a generalisation of the ideas we met in the previous section but now allowing for the presence of different particle types, some of which may cross the membrane. Let us now make the assumption that there is no interaction between the particle types 1 and 2 in B. We shall comment on this shortly. With this assumption, we may say that the behaviour of particles 1 in B is just as if particles 2 are absent and we may treat particles 1 as if they alone occupy the whole system. You are asked to show in question 3, Chapter 10, Appendix 4 that the *equality of the chemical potentials* for particles of type 1 on either side of the partition means that the pressures exerted by this type of particle are equal *because they exist in the same phase*, here a gas i.e. $P_A{}^1 = P_B{}^1$. If we now add the pressure exerted on the right-hand side by the gas 2, we see that there is an excess pressure of $P_B{}^2$ in the right-hand volume: this is the osmotic pressure. Although we have chosen to frame our discussion in terms of gases, we could have considered 1 to be a liquid and 2 to be another liquid, or as is usually the case, a solute dissolved in the liquid acting as a solvent. Osmosis is of vital importance in the functioning of living organisms. In a cell, the osmotic pressure is balanced by the stresses in the cell wall.

Let us now comment on our assumption of non-interaction. If our particles are ideal gases, this is certainly so. If, however, the particles are liquids, or a liquid and a solute, this assumption is not valid but it acts as a first approximation. Correction terms have then to be added to the osmotic pressure to take account of these interactions but a study of these would take us further than we would wish to go.

The Fermi level

In the free electron theory of metals, the electrons are treated as a gas in which the electron particles obey quantum statistics. It is found that, at absolute zero, the electrons completely fill a range of energy levels up to a maximum energy called the *Fermi level*. At higher temperatures this maximum energy differs only very slightly from the Fermi energy. Any additional electron added to the metal has to be added at the Fermi level because all the lower energy levels are already occupied. This is represented schematically in the energy level diagram given in Fig. 10.3. The work function ϵ_W is the energy required to liberate an electron, at the Fermi level, from the metal.

We know from equation [10.2] that μ is the increase in the internal energy upon the addition of one particle, under condi-

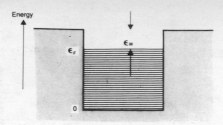

Fig. 10.3 A schematic representation of the energy levels of the electrons in a metal.

tions of constant entropy and volume. These conditions mean that the system is isolated from the surroundings apart from the reversible addition of the particle: the constant V implies that no work is done on the system and the constant S implies that no heat is added. Now when an electron is added to the metal, it enters at the Fermi level and so the energy of the metal is then increased by ϵ_F. Thus we can identify μ with ϵ_F.

Now consider two different metals in contact. Each will have its own Fermi energy and work function. When they are put in contact so that there can be electron flow between them, the chemical potentials must be equal at equilibrium. This means that the Fermi levels are the same and we have the energy level diagram as in Fig. 10.4. It can be seen from this that there will be a contact potential difference between them equal to $(\epsilon_{W_2} - \epsilon_{W_1})/e$ where e is the electron charge. This idea can be extended to semiconductors and is of fundamental importance in the operation of junction diodes.

The condition for chemical equilibrium

The chemical potential is aptly named because it is of particular importance in chemistry. As just one example of its use there, we shall derive the general condition for chemical equilibrium in a reaction.

Let us consider the reaction

$$H_2 + Cl_2 = 2HCl$$

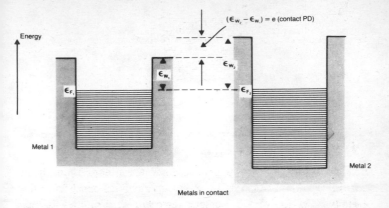

Fig. 10.4 The physical origin of the contact potential. When two metals are put into electrical contact, their chemical potentials are equal at equilibrium. This means that their Fermi levels are the same as shown. The contact potential difference is equal to the difference in the work functions $(\epsilon_{W_2} - \epsilon_{W_1})$ divided by the electron charge.

which can be written as

$$2HCl - H_2 - Cl_2 = 0 \qquad [10.26]$$

A general way of writing such a reaction is

$$\sum_i \nu_i A_i = 0 \qquad [10.27]$$

where the A_i denote chemical symbols and the so-called *stoichiometric coefficients* ν_i are either positive or negative small integers. We shall adopt the convention, assuming that the reaction proceeds in a given direction, that ν_i is positive if a molecule is formed in the reaction and is negative if one 'disappears' as a reactant. In the example that we have just given, $\nu_{H_2} = \nu_{Cl_2} = -1$ and $\nu_{HCl} = 2$.

Let now N_i be the number of molecules of type i involved in the reaction. These numbers N_i will change as the reaction proceeds, but they cannot change independently of each other.

The numbers can change only in a way which is consistent with the equation denoting the chemical reaction because the numbers of the different types of atoms are conserved. For each molecule of H_2 and each molecule of Cl_2 that disappear upon reaction, two new molecules of HCl appear. The change in the number N_i must therefore be proportional to the stoichiometric coefficients appearing in equation [10.27] for the chemical reaction. Thus we may write

$$dN_i = \lambda \nu_i \qquad [10.28]$$

where the constant λ is the same for all the different types of molecules involved in the reaction. In our example,

$$dN_{HCl} : dN_{H_2} : dN_{Cl_2} = 2 : -1 : -1$$

Now if a reaction is open to the surrounding atmosphere where the pressure and temperature are fixed, we know from our discussion in section 6.4 that G will be a minimum at equilibrium. In any infinitesimal process then at equilibrium, $dG = 0$ and so equation [10.16] gives

$$dG = \sum_i \mu_i dN_i = 0 \qquad [10.29]$$

Using equation [10.26], this becomes

$$\boxed{\sum_i \mu_i \nu_i = 0} \qquad [10.30]$$

This is the general condition for chemical equilibrium in a reaction. *It tells us that chemical potentials are additive in the same proportions in which the chemical components occur in the reaction.* For example, in the reaction for the formulation of HCl considered at the beginning of this section,

$$\mu_{HCl} = 1/2 \, (\mu_{Cl_2} + \mu_{H_2})$$

which is a useful result.

Chapter 11
The third law of thermodynamics

11.1 The Nernst heat theorem

The third law of thermodynamics is concerned with the entropy of a system as the temperature is reduced to absolute zero. If we integrate equation [5.4] from absolute zero to a temperature T,

$$S = \int_0^T \text{đ}Q/T + S_0 \qquad [11.1]$$

but we cannot determine S_0, the entropy at absolute zero, in any way from the second law. The third law however enables us to give a value to S_0.

The original statement of the third law was given by Nernst in 1906. Nernst noticed that, in many chemical reactions occurring with no change in the end point temperatures, the values of ΔG decreased while that of ΔH increased. Nernst postulated as his heat theorem that not only did these two quantities become equal at $T = 0$ but they also approached each other asymptotically, as in Fig. 11.1. We shall now see what this means in terms of the entropy. It follows from equation [6.37] that

$$\Delta G = \Delta H - T\Delta S \qquad [11.2]$$

for our chemical reaction and this clearly shows that $\Delta G \rightarrow \Delta H$ as $T \rightarrow 0$. However, in order for the curves to touch each other asymptotically as in Fig. 11.1, it may be shown that ΔS itself must vanish as $T \rightarrow 0$. These ideas are embodied in the following rather more general statement of the Nernst heat theorem than was originally given by Nernst:

The entropy change in a process, between a pair of equilibrium states, associated with a change in the external parameters tends to zero as the temperature approaches absolute zero.

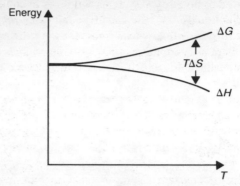

Fig. 11.1 Nernst postulated that the curves for ΔH and ΔG in a chemical reaction approach each other asymptotically as $T \rightarrow 0$.

By external parameters, we mean such variables as pressure, temperature, magnetic field and so on. Ever since its formulation there has been a great deal of discussion about the significance of the Nernst heat theorem, but there is now a great deal of experimental evidence in its support and so it has assumed the status of a fundamental law — the third law of thermodynamics.

The original Nernst formulation of the third law has subsequently been followed by various other statements which are more in accord with our understanding of modern quantum statistical mechanics. A particularly useful statement of the third law was given by Planck in 1911. It is a somewhat more powerful statement than the earlier statement of Nernst and is:

The entropy of all perfect crystals is the same as the absolute zero, and may be taken as zero.

By a perfect crystal we mean one in which the arrangement of the atoms repeats itself on a regular basis throughout the crystal. In a glass, on the other hand, there is no such regular repetition or, as we say, no long-range order. The essential point in the Planck statement is that the entropies of such crystals are equal at $T = 0$; it is then a matter of convenience to put $S_0 = 0$. This is a sensible choice because it gives agreement with the microscopic view to be discussed in the next section. Athough the Planck statement is usually quoted for perfect crystals for historical reasons, it is now believed to hold for all systems which are in equilibrium states, including liquids (e.g. ^3He and ^4He) and gases. It might seem somewhat surprising that we can consider a gas existing at absolute zero but there are certain quantum systems, such as the electrons in a metal, which do constitute a gas-like assembly even down to $T = 0$.

Let us remind ourselves what we mean by an equilibrium state from an energy point of view and of the difference between *stable* and *metastable* equilibrium. Under given conditions, a system in equilibrium is in a state corresponding to a minimum in the appropriate potential function. For example a complex system at constant P and T takes a minimum value of G at equilibrium when another macroscopic variable, such as the volume, is varied. Such a state is one of stable equilibrium because *any* departure from this state entails an increase in G. If on the other hand G increases for *small departures only* from the equilibrium state before decreasing again for larger departures, we say that the state is one of metastable equilibrium. These ideas allow us to test the validity of the third law in a very beautiful way as follows.

Many substances exist in different allotropic forms. We give two examples.
1. Grey and white tin. Above 291 K the stable form of tin is white tin, with a tetragonal crystal structure, while below this temperature the stable form is cubic grey tin.
2. Monoclinic and rhombic sulphur. Above 367 K the stable form of sulphur is monoclinic sulphur while below this temperature the stable form is rhombic sulphur.

Consider now one such substance which can exist in two allotropic ordered forms. At a given temperature each form can exist in an equilibrium state with a minimum in the energy, although the more stable form will have the lower energy minimum. The form with the higher energy minimum is in metastable equilibrium. If sufficient energy is given to the system in this metastable equilibrium state, so that the potential barrier between the states can be overcome, the system will change to the lower energy state of stable equilibrium. This is analogous to a ball being kicked out of a hollow so that it rolls down to the bottom of a hill. At high temperatures, where $k_B T$ is larger than or comparable to the height of the potential barrier, there is sufficient random thermal vibrational energy to induce such a change; thus a system initially in the metastable state will change gradually into the stable state. We cannot regard a system in a metastable state at high temperatures as being in a *true* equilibrium state because the state *changes with time*. However at low temperatures, and certainly at the absolute zero, we can regard such a metastable state as being a true equilibrium state because there is now insufficient thermal energy to induce a change. Our conclusion is that the Planck form of the third law may be applied to the system in either the stable or the metastable equilibrium states.

As a specific application of these ideas, let us consider sulphur. If monoclinic sulphur is cooled rapidly through 367 K down to a very low temperature, the monoclinic phase is locked in, with the transition rate to the more stable rhombic form becoming negligible. Thus, at low temperatures both forms of sulphur can be produced in stable equilibrium states with S_0 being the same for both forms. If the Planck statement of the third law holds, then the entropy must vary for the two forms as in Fig. 11.2, with a common value of S_0. In particular, the entropy change $S_B - S_0$ in monoclinic sulphur between 0 K and 367 K is given by the two expressions

$$S_B - S_0 = \int_0^{367} \frac{C_P^{\text{monoclinic}}}{T} \, dT$$

and $\quad S_B - S_0 = S_B - S_A + S_A - S_0 = \dfrac{l}{367}$

$$+ \int_0^{367} \frac{C_P^{\text{rhombic}}}{T} \, dT$$

where l is the latent heat for the transition. Now the heat capacities for each of the two forms can be measured from close to absolute zero up to 367 K. Also l can be measured and so we can find the two values for $S_B - S_0$. Typical experimental values are 37.82 ± 0.40 and 37.95 ± 0.20 kJ K^{-1} mol^{-1}. As these values are the same to within the experimental error, we conclude that our assumption that S_0 is the same for each form of sulphur is valid. This is also found for other different allotropes which justifies our confidence in the Planck statement of the third law.

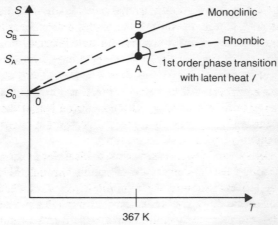

Fig. 11.2 The entropies of monoclinic and rhombic sulphur as a function of temperature. These curves may be obtained from a measurement of the temperature dependence of the heat capacity for each phase. We can deduce from these experimental measurements, together with a measurement of the latent heat for the first-order phase transition at 367 K, that the entropy curves meet each other at $T = 0$. This is consistent with the third law.

Although there are some systems which appear to violate the third law, in that they have a non-zero entropy at the absolute zero, there is in fact never any violation because such systems are not in *true equilibrium states*. The most important examples of such systems are glasses and we shall discuss them at the end of section 11.3.

11.2 A microscopic viewpoint of the third law

It is very instructive to see if we can understand the origins of the third law from a microscopic viewpoint. We know that, according to quantum mechanics, the small particles comprising the system can exist in different quantum states with discrete energies such as those shown in Fig. 6.3. As the temperature is lowered, the particles all pack into the lowest energy levels until, at absolute zero, they are all in the ground state (supposing for the moment that they are bosons). Let us also suppose that the ground state is non-degenerate with $g = 1$. This is a very controversial assumption as it is not usually true, but we shall make it nevertheless because the result that it gives is illuminating. With these assumptions we see that the number of different ways Ω in which all the particles can be arranged in the ground state is 1. An application of

$$S = k_B \ln \Omega \qquad [5.12]$$

shows that $S_0 = 0$ at the absolute zero which is the Planck statement. This of course also implies that $\Delta S = 0$ for any process occurring at the absolute zero, which is precisely the Nernst heat theorem.

We should now comment on this simple but instructive argument for the decrease of the entropy to zero at the absolute zero. The argument hinged on the discreteness of the energy levels, as well as on the non-degeneracy of the ground state. However, for the macroscopic systems of concern in thermodynamics, these levels are so closely spaced that, at temperatures of a few degrees K where the measured decrease in S from the room temperature value is quite marked, there are so many states which may be occupied that we cannot say the system is in *one* lowest energy state with $\Omega = 1$. The explanation is that the decrease in entropy towards zero depends in fact on the behaviour of the number of states per unit energy range (the *density of states*) with energy, rather than just on the occupancy of the ground level as our simple theory suggests. In particular, it is the behaviour of the density of states at low energies which determines the low temperature properties so that the entropy certainly does fall to zero as T tends to 0 for both bosons and fermions. We shall not pursue this point further as this would take us too far into the realm of statistical mechanics.

11.3 The Simon formulation of the third law

Let us for a moment consider cooling our perfect crystal down towards absolute zero. As we cool the crystal, the lattice vibrations reduce and we begin to achieve a state of perfect order, with the atoms all settling in their regularly arranged positions and S tending towards zero. However we know that the electrons in the atoms can have a net electron spin and these can have different energies according to whether they point along or against an applied magnetic field. There will thus be a spin entropy remaining, due to the different orientations of the electron spins, when all the lattice vibrational entropy has disappeared. Let us cool the crystal even more so that the spins all go into the same ground state, this being the most ordered arrangement with zero spin entropy. Do we now have zero entropy for the system? The answer is almost certainly no as we have forgotten about the weak nuclear spins, which again can be distributed amongst a set of very closely-spaced energy levels. We have to reduce the temperature even further to remove this nuclear entropy. Of course this argument could be pursued even further by considering other contributions to the entropy within the nucleons and so on. In our second statement of the third law we covered ourselves against these hidden entropies by stipulating a perfect (and by implication simple) crystal where there are no electron or nuclear spins, only atomic size masses in a regular arrangement.

The important point is that the lattice, the electron spin system and the nuclear system are essentially uncoupled from each other at low enough temperatures and act as independent systems, each in internal thermodynamic equilibrium. Their entropies are additive. Simon called these independent systems *aspects* of the whole system and gave in 1937 the following general statement of the third law:

> *The contribution to the entropy of a system from each aspect which is in internal thermodynamic equilibrium disappears at absolute zero.*

The Simon statement is convenient because it means that we can focus our attention on just one aspect of interest, with the knowledge that its entropy is zero at $T = 0$.

We conclude this section with a brief comment about glasses. Many liquids where the disorder, and consequently the entropy, are high, retain their liquid structure if they are cooled rapidly through their freezing point to form a glass, with frozen-in entropy at absolute zero. In contrast, if they are cooled slowly, they go into the usual ordered crystal phase, with zero entropy at absolute zero. The crystal phase is the stable phase, with a minimum in the energy, while the glass is an unstable phase, not being at an energy minimum. The glass will slowly crystallise, although the time period for this may be years or even considerably longer. The glass phase is thus *not an equilibrium state*. Glycerine, with a melting point at 19 °C, is a good example of such a material. Although the frozen-in entropy persists down to absolute zero, this in no way violates the third law because this law relates only to systems in equilibrium.

11.4 Some consequences of the third law

It is consequent on the third law that certain measurable parameters vanish at the absolute zero of temperature. Let us examine these in turn.

The thermal expansion coefficient

We have

$$\beta = \frac{1}{V} \left(\frac{\partial V}{\partial T} \right)_P \qquad [2.1]$$

This can be transformed by the Maxwell relation

$$\left(\frac{\partial V}{\partial T} \right)_P = - \left(\frac{\partial S}{\partial P} \right)_T \qquad [6.41]$$

to

$$\beta = - \frac{1}{V} \left(\frac{\partial S}{\partial P} \right)_T \qquad [11.3]$$

But we know by our Nernst formulation of the third law that the entropy change in an isothermal process tends to zero at the absolute zero, so the partial differential in equation [11.3] is zero. We conclude that *the thermal expansion coefficient is zero at absolute zero.*

The temperature dependence of the magnetic moment in a magnetic system

An application of our mnemonic shows that there is a Maxwell relation

$$\left(\frac{\partial \mathcal{M}}{\partial T}\right)_{B_0} = \left(\frac{\partial S}{\partial B_0}\right)_T \qquad [11.4]$$

and the right-hand side vanishes at $T = 0$ by the third law, as in the previous section. Hence $(\partial \mathcal{M}/\partial T)_{B_0} = 0$ at absolute zero, which means that there is no temperature dependence of the magnetic moment there. This immediately tells us that the Curie law cannot hold down to absolute zero, as the following argument shows.

Suppose the Curie law is obeyed as

$$\chi = \frac{\mathcal{C}}{T} = \frac{\mu_0 \cdot \mathcal{M}}{VB_0} \qquad [11.5]$$

Thus $\left(\dfrac{\partial \mathcal{M}}{\partial T}\right)_{B_0} = -\dfrac{VB_0 \, \mathcal{C}}{\mu_0 T^2}$ $\qquad [11.6]$

and this clearly does not vanish as $T \to 0$. We conclude then that *the Curie law breaks down at very low temperatures.*

The physical reason for the breakdown of Curie's law in a magnetic system is that there is always an interaction between the elementary magnetic dipoles. In a magnetic salt this interaction is usually very weak and is negligible compared with the thermal energy $k_B T$ at high temperatures; as a result, Curie's law holds. At low temperatures this interaction becomes important, so the salt departs from Curie's law, with some sort of magnetic ordering occurring and the salt becoming usually ferro- or antiferromagnetic. The magnetic moment then does not change with temperature at $T = 0$, the conclusion we reached above.

The heat capacity

Let us for the moment consider C_V. We have

$$C_V = T \left(\frac{\partial S}{\partial T} \right)_V \qquad [6.9]$$

or

$$C_V = \left(\frac{\partial S}{\partial \ln T} \right)_V \qquad [11.7]$$

As $T \to 0$, $\Delta S \to 0$ by the Nernst statement of the third law. But $\Delta(\ln T)$ is certainly non-zero as $T \to 0$ because $\ln T \to -\infty$ there. Hence C_V *should tend to zero as* $T \to 0$. A similar argument holds for the other heat capacities.

This is always observed experimentally. For example, the heat capacity of most metals at low temperatures is found to obey the law

$$C_P = aT + bT^3 \qquad [11.8]$$

down to the very lowest temperatures attainable. The first term is due to the conduction electrons and the second is the contribution from the lattice. This heat capacity vanishes at $T = 0$ in agreement with the third law.

There is a more instructive argument that we can give for the vanishing of the heat capacities at the absolute zero. From equation [6.9], considering C_V for example,

$$S - S_0 = \int_0^T \frac{C_V}{T} \, dT$$

Now the left-hand side of this equation must remain finite at all temperatures down to absolute zero, where it vanishes by the third law, so the right-hand side must remain finite also. This means that C_V must fall to zero as $T \to 0$, at least as fast as T; otherwise the right-hand side would diverge there. Equation [11.8] is consistent with this idea.

The slope of the phase boundary in a first-order transition

The slope of the phase boundary in a *first*-order transition is given by the Clausius–Clapeyron equation

$$\frac{\mathrm{d}P}{\mathrm{d}T} = \frac{\Delta S}{\Delta V} \qquad\qquad\qquad [9.8]$$

As ΔS tends to zero as $T \to 0$ by the Nernst statement of the third law, we see that *the slope of a phase boundary in a first-order transition is zero at the absolute zero*. This is shown in Figs. 9.12 and 9.13, the phase diagrams for ^{4}He and ^{3}He. Another nice application of this idea is in superconductors. It can be shown in exactly the same way that the phase boundary between the normal and the superconducting phases has zero slope at $T = 0$; this is illustrated in Fig. 9.10.

> *The fact that all these predictions based on the third law are in agreement with experimental observations may be taken as the experimental confirmation of the third law.*

11.5 The unattainability of absolute zero

There is yet another statement of the third law:

> *It is impossible to reach absolute zero using a finite number of processes.*

Although there are formal proofs to show that this statement of the third law is equivalent to the Nernst statement (see for example the text by Zemansky), we shall not give such a proof here. Instead, we shall resort to a physical argument to show that the two statements are consistent with each other.

We have seen in the previous chapter that the lowest temperatures attainable experimentally are achieved using the adiabatic demagnetisation technique. Let us consider a whole series of successive demagnetisations in an attempt to reach absolute zero, as illustrated in Fig. 11.3.

In Fig. 11.3 we have drawn two entropy curves for a magnetic salt in zero applied induction field and in a finite field B_0. As the elementary dipoles are aligned by the field, they are in a state of greater order than when they are unaligned in the absence of the external field. Thus the entropy curve for the state in the applied field is the lower one. At absolute zero, by the Nernst statement of the third law, the difference in entropy between the two curves is zero, and this is shown in Fig. 11.3(a). In Fig. 11.3(b) we have drawn the two entropy curves in a way that violates the third law.

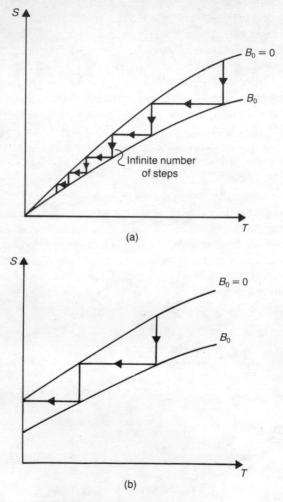

(a)

(b)

Fig. 11.3 It is impossible to reach absolute zero in a finite number of steps. The process illustrated here is for the adiabatic demagnetisation of a paramagnetic salt. (a) If the third law is true. (b) If a third law is untrue.

Now consider the series of isothermal magnetisations and adiabatic demagnetisations as represented by the zig-zag paths in Fig. 11.3. Each successive demagnetisation reduces the temperature. If the entropy curves were as in Fig. 11.3(b), then absolute zero could be reached in a finite number of operations. However we know that the entropy curves are as in Fig. 11.3(a), and it is clear that absolute zero cannot be obtained in a finite number of demagnetisations, an infinite number being required. We thus conclude that it is impossible to reach absolute zero in any practical way. Nevertheless, much progress has been made in recent years to reach extremely low temperatures. In section 9.11 the discovery of the superfluid phases of ^{3}He, at the incredibly low temperatures of a few millikelvin was described. Since then several laboratories have achieved lattice temperatures of a few microkelvin while, at one laboratory in Helsinki, the nuclei of copper have been cooled at a few nanokelvin. At these temperatures they form a nuclear antiferromagnet. The absolute zero will always be unattainable, but we can be confident that a wealth of fascinating physics remains to be discovered as our techniques improve and we are able to approach ever and ever closer to this elusive end.

Appendix 1

Values of physical constants and conversion factors

Quantity	Symbol	Value
Gas constant	R	$8.31 \, \text{J K}^{-1} \text{mol}^{-1}$
Avogadro constant	N_A	$6.02 \times 10^{23} \text{mol}^{-1}$
Boltzmann constant	k_B	$1.38 \times 10^{-23} \text{J K}^{-1}$
Stefan constant	σ	$5.67 \times 10^{-8} \text{W m}^{-2} \text{K}^{-4}$
Planck constant	h	$6.63 \times 10^{-34} \text{J s}$
Magnitude of the electron charge	e	$1.60 \times 10^{-19} \text{C}$
Faraday constant	$F_0 = e N_A$	$96\,485 \text{ C}$
Speed of light	c	$3.00 \times 10^8 \text{m s}^{-1}$
Acceleration due to gravity	g	9.81 m s^{-2}
Permeability of free space	μ_0	$4\pi \times 10^{-7} \text{H m}^{-1}$
Permittivity of free space	ϵ_0	$8.85 \times 10^{-12} \text{F m}^{-1}$
Mechanical equivalent of heat	J	$4.19 \text{ J calorie}^{-1}$
Molar volume of an ideal gas at STP		22.4 litres
Atmospheric pressure		$1.01 \times 10^5 \text{N m}^{-2}$ $= 760 \text{ mm of Hg}$ $= 760 \text{ torr}$
1 horsepower	hp	746 W
1 kilowatt hour	kWh	$10^3 \times 60 \times 60 \text{ J}$ $= 3.6 \times 10^6 \text{J}$

Appendix 2

Some mathematical relations used in thermodynamics

The reciprocal and the cyclical relations

Suppose that there exists a relation between the variables x, y and z

$$F(x, y, z) = 0 \qquad \text{[A2.1]}$$

so that only two of them are independent. We could rearrange equation [A2.1] to give x as a function of y and z as

$$x = x(y, z) \qquad \text{[A2.2]}$$

where $x(y, z)$ stands as usual for a function of y and z. Now we know that the infinitesimal change dx in x consequent on the infinitesimal changes dy and dz in y and z is

$$dx = \left(\frac{\partial x}{\partial y}\right)_z dy + \left(\frac{\partial x}{\partial z}\right)_y dz \qquad \text{[A2.3]}$$

Similarly, writing $y = y(x, z)$

$$dy = \left(\frac{\partial y}{\partial x}\right)_z dx + \left(\frac{\partial y}{\partial z}\right)_x dz \qquad \text{[A2.4]}$$

If we now substitute for dy from equation [A2.4] in equation [A2.3]

$$dx = \left(\frac{\partial x}{\partial y}\right)_z \left(\frac{\partial y}{\partial x}\right)_z dx + \left[\left(\frac{\partial x}{\partial y}\right)_z \left(\frac{\partial y}{\partial z}\right)_x + \left(\frac{\partial x}{\partial z}\right)_y\right] dz$$
$$\text{[A2.5]}$$

Let us now choose x and z to be the independent variables. This means that we can have $dz = 0$ in equation [A2.5] and still have a non-zero value for dx, as x and z are independent. With this value of dz and with the common term dx cancelled, equation [A2.5] becomes

$$1 = \left(\frac{\partial x}{\partial y}\right)_z \left(\frac{\partial y}{\partial x}\right)_z$$

or

$$\boxed{\left(\frac{\partial x}{\partial y}\right)_z = \left(\frac{\partial y}{\partial x}\right)_z^{-1}} \qquad [A2.6]$$

This is known as the *reciprocal relation*.

Alternatively, we could choose d$x = 0$ with d$z \neq 0$ in equation [A2.5] yielding

$$\left(\frac{\partial x}{\partial y}\right)_z \left(\frac{\partial y}{\partial z}\right)_x + \left(\frac{\partial x}{\partial z}\right)_y = 0$$

or

$$\boxed{\left(\frac{\partial x}{\partial y}\right)_z \left(\frac{\partial z}{\partial x}\right)_y \left(\frac{\partial y}{\partial z}\right)_x = -1} \qquad [A2.7]$$

using equation [A2.6]. This is known as the *cyclical relation* or, more usually, the *cyclical rule*. It is easy to remember because of the cyclical order. Note the -1 on the right-hand side.

The chain rule

Suppose again that x, y and z are not independent, being related by equation [A2.1]. Let us consider some function ϕ of x, y and z. Because of equation [A2.1], ϕ may be expressed in terms of only two of the variables, say

$$\phi = \phi(x, y) \qquad [A2.8]$$

Equation [A2.8] can be rearranged to give

$$x = x(\phi, y) \qquad [A2.9]$$

so

$$dx = \left(\frac{\partial x}{\partial \phi}\right)_y d\phi + \left(\frac{\partial x}{\partial y}\right)_\phi dy \qquad [A2.10]$$

Dividing equation [A2.10] all through by dz, holding ϕ constant,

$$\left(\frac{\partial x}{\partial z}\right)_\phi = \left(\frac{\partial x}{\partial y}\right)_\phi \left(\frac{\partial y}{\partial z}\right)_\phi \qquad [A2.11]$$

This is the *chain rule*. We should note the common ϕ outside each partial differential. The chain rule must not be confused with the cyclical rule, which is a relation just between the variables x, y and z with no other function ϕ being involved.

The condition for a differential to be exact

A mathematical function $\phi\,(x,y)$ of x and y takes unique values for each pair of values of x and y. When x and y change by dx and dy, the infinitesimal change in ϕ is

$$d\phi = \left(\frac{\partial \phi}{\partial x}\right)_y dx + \left(\frac{\partial \phi}{\partial y}\right)_x dy \qquad [A2.12]$$

Because it is the differential of a mathematical function, dϕ is called *an exact differential*. A finite change in ϕ when x changes from x_1 to x_2 and y from y_1 to y_2 is

$$\Delta\phi = \phi(x_2,y_2) - \phi(x_1,y_1) = \int_{x_1 y_1}^{x_2 y_2} d\phi \qquad [A2.13]$$

As the values of ϕ at the points (x_1,y_1) and (x_2,y_2) are fixed, then $\Delta\phi$ is also fixed; consequently, it does not matter how we vary x and y during the integration between the given limits. We say that the integral is *path independent*. In thermodynamics it is frequently important to know whether an integral is path independent; in other words, we have to establish whether the integrand is an exact differential. There is a simple test for this.

Suppose we encounter a differential of the form

$$dG = X\,dx + Y\,dy \qquad [A2.14]$$

where X and Y are in general functions of both x and y. We wish to establish now whether dG is exact. If we differentiate the co-efficient of dx in equation [A2.12] with respect to y, holding x

constant, we obtain $\partial^2\phi/\partial y \partial x$, while differentiating the coefficient of dy with respect to x, holding y constant, gives $\partial^2\phi/\partial x \partial y$. Now it is shown in the standard texts on partial differentiation that these two partial differentials are equal, the order of the differentiation being immaterial. If dG is an exact differential, then

$$\left(\frac{\partial X}{\partial y}\right)_x = \left(\frac{\partial Y}{\partial x}\right)_y \qquad \text{[A2.15]}$$

Our argument shows that equation [A2.15] is a *necessary* condition for dG to be exact; it may also be shown, using a more sophisticated argument, to be *sufficient*.

As an example of this consider the differential

$$dG = 2xy^4\,dx + 4x^2y^3\,dy \qquad \text{[A2.16]}$$

We see that the test equation [A2.15], is satisfied by the differential dG given in equation [A2.16] and so we conclude that it is an exact differential. In fact the actual function G is yielded immediately upon integration. From the form of dG in equation [A2.16] we have

$$\left(\frac{\partial G}{\partial x}\right)_y = 2xy^4 \quad \text{so} \quad G(x,y) = x^2y^4 + f(y)$$

$$\left(\frac{\partial G}{\partial y}\right)_x = 4x^2y^3 \quad \text{so} \quad G(x,y) = x^2y^4 + g(x)$$

The only way for these two solutions for G to be equal is for $f(y) = g(x) = $ a constant. Thus $G(x,y) = x^2y^4 + $ a constant. On the other hand, it may be instantly established that the differential

$$đG = xy^4\,dx + 4x^2y^3\,dy$$

is inexact.

Appendix 3

The work required to magnetise a magnetic material and to polarise a dielectric

Magnetic work

Consider a sample of magnetic material becoming magnetised by being placed inside a long solenoid as in Fig. A3.1. The length of the sample is l, the cross-sectional area is A and it fits exactly inside the whole volume of the solenoid. The current is quasistatically increased, and with it the applied magnetisation.

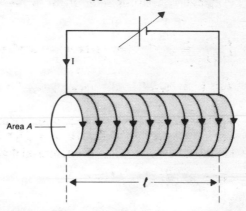

Fig. A3.1 A magnetic material contained within a solenoid. The current is gradually increased from zero so that the material is magnetised.

We have the following relations:

$$B = \mu_0(H + M) = B_0 + \mu_0 M$$

where B = magnetic induction, H = applied magnetising field or, more commonly, the magnetic field, M = magnetisation or the magnetic moment per unit volume, B_0 is the induction in the *absence* of the specimen and μ_0 = permeability of free space.

$$M = \chi_m H$$

for a linear magnetisable material such as a paramagnet where χ_m is the magnetic susceptibility.

$$B = \mu_0(1 + \chi_m) H = \mu\mu_0 H$$

where $\mu = 1 + \chi_m$ is the permeability.

$$. \mathscr{M} = VM$$

where $. \mathscr{M}$ is the overall magnetic movement.

We are assuming, for simplicity, that all the vector quantities in the above relations are parallel, and so may treat them as scalars. Also, we shall consider them as uniform over the volume V of the long solenoid, there being no significant end effects.

We know that the magnetic field in the middle of a long solenoid, carrying a current I and with n turns per unit length, is $H = nI$. This means that

$$B_0 = \mu_0 nI$$

The flux threading the solenoid is:

$$\Phi = BAnl = BnV$$

If the current is increased from its instantaneous value I to $I + \mathrm{d}I$ in a time $\mathrm{d}t$, there is a back EMF.

$$\mathscr{E} = nV\frac{\mathrm{d}B}{\mathrm{d}t}$$

and it is the battery having to drive charge around the circuit against this back EMF that is the source of the work required to magnetise the sample.

The work done by the battery in the time $\mathrm{d}t$, when charge $I \, \mathrm{d}t$ flows, is

$$đW = \mathscr{E} I\mathrm{d}t = nV\frac{\mathrm{d}B}{\mathrm{d}t} I\mathrm{d}t = \frac{B_0 V}{\mu_0} \mathrm{d}B$$

or $\quad \text{đ}W = \dfrac{B_0 V}{\mu_0}\left[\mathrm{d}B_0 + \mu_0 \mathrm{d}M \right]$

So the total work done in the magnetisation process is

$$W = V \int \dfrac{B_0 \mathrm{d}B_0}{\mu_0} + V \int B_0 \mathrm{d}M = V \int \dfrac{B_0 \mathrm{d}B_0}{\mu_0} + \int B_0 \mathrm{d}.\, \mathscr{M}$$

There are two terms here. The first is just the familiar energy term that the solenoid would have in the absence of the magnetic sample; upon integration, it gives the energy density $(B_0{}^2)/(2\mu_0)$. The second is the work required to bring the sample up to its final magnetisation. We conclude that the infinitesimal work required to increase the overall magnetic moment from $\mathscr{M}$ to $\mathscr{M} + \mathrm{d}.\,\mathscr{M}$ in the applied induction B_0 is

$$\boxed{\text{đ}W = B_0 \mathrm{d}.\, \mathscr{M}}$$

If the magnetisation and the magnetic field are not constant over the volume of the sample as we have assumed, this argument may be extended to give

$$\text{đ}W = \int B_0 \mathrm{d}M \mathrm{d}V,$$

where the integration takes place over the whole volume of the sample.

Dielectric work

Let us now consider quasistatically polarising a dielectric by placing it between the plates of a parallel plate condenser as in Fig. A3.2 and gradually increasing the voltage $\mathscr{V}$ across the plates. We shall assume that the dielectric exactly fills the space between the plates and we shall assume a uniform field there by neglecting any edge effects.

The following relations hold:

$$D = \epsilon_0 E + P$$
$$P = \epsilon_0 \chi_e E$$
$$D = \epsilon_0 (1 + \chi_e)\, E = \epsilon_0 \epsilon E$$
$$\mathscr{P} = VP$$

where E = electric field, D = electric displacement, P = polarisation or the electric dipole moment per unit volume, ϵ_0 = permittivity of free space, $\epsilon = 1 + \chi_e$ = dielectric constant, χ_e = electric susceptibility, P = overall electric dipole moment.

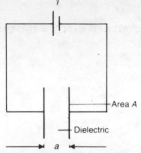

Fig. A3.2 A dielectric material between the plates of a parallel plate condenser. The voltage across the condenser is gradually increased from zero so that the material is polarised.

As for the magnetic case, we are assuming that all the vector quantities in the above relations are parallel, and so may treat them as scalars.

If the free charge density on the plates is σ, Gauss' law gives

$$D = \sigma = Z/A$$

where Z is the charge on each plate. Now increase the charge by dZ. The battery has to do the work

$$\begin{aligned} đW &= VdZ = (aE)(A\,dD) = VE\,dD \\ &= V(\epsilon_0 dE + dP)E \end{aligned}$$

where the volume $V = aA$. Integrating,

$$W = \epsilon_0 V \int E\,dE + V \int E\,dP$$

The first term is the familiar energy term for the energy of an empty charged capacitor with an energy density $\epsilon_0 E^2/2$ between the plates. The second term is then the work done in polarising the dielectric. We conclude that the infinitesimal work required to increase the overall dipole moment of a dielectric from $\mathscr{P}$ to $\mathscr{P} + d\mathscr{P}$ in the field of E is

$$\boxed{đW = E\,d\mathscr{P}}$$

If the polarisation and the electric field are not constant over the volume of the sample as we have assumed, this argument may be extended to give

$$\text{d}W = \int E \text{d}P \text{d}V$$

where the integration is over the volume of the dielectric.

Appendix 4
Questions

Chapter 1

1   The length of the mercury column in a mercury-in-glass thermometer is 5 cm when the bulb is immersed in water at its triple
point. What is the temperature on the mercury-in-glass scale when
the length of the column is 6.0 cm? What will the length of the
column be when the bulb is immersed in a liquid at 100 degrees,
as measured on the mercury-in-glass scale, above the ice point? If
the length of the column can be measured to within only 0.01 cm,
can this thermometer be used to distinguish between the ice point
and the triple point of water? You may take the temperature of
the ice point, as measured on the mercury-in-glass scale, as 273.15
degrees.

2   The resistance of a wire is given by

$$R = R_0(1 + \alpha t + \beta t^2) \qquad \text{341.79}°$$

where t is the temperature in degrees Celsius measured on the
ideal gas scale and so R_0 is the resistance at the ice point. The
constants α and β are $3.8 \times 10^{-3} \text{ K}^{-1}$ and $-3.0 \times 10^{-6} \text{ K}^{-2}$
respectively. Calculate the temperature on the resistance scale at a
temperature of $70\,°C$ on the ideal gas scale.

3   The table below lists the observed values of the pressure P of a gas
in a constant-volume gas thermometer at an unknown temperature
and at the triple point of water as the mass of gas used is reduced.

P_{TP} (torr)	100	200	300	400
P (torr)	127.9	256.5	385.8	516

By considering the limit (P/P_{TP}) determine T to two decimal
$$P_{TP} \to 0$$

places. What is this in $°C$? (1 torr is a pressure of 1 mm of Hg.)

4 How many kilograms of helium gas are contained in a chamber of 1 litre volume at $50\,°C$ if the pressure is one atmosphere? (The atomic weight of He is 4.)

5 A mixture of hydrogen and oxygen is isolated and allowed to reach a state of constant pressure and temperature. The mixture is exploded with a spark of negligible energy and is allowed to reach a state of constant temperature and pressure again. Is the initial state an equilibrium state? Is the final state an equilibrium state?

Chapter 2

1 10 moles of an ideal gas are compressed isothermally and reversibly from a pressure of 1 atmosphere to 10 atmospheres at 300 K. How much work is done?

2 An ideal gas undergoes the following reversible cycle: (a) an isobaric expansion from the state (P_1, V_1) to the state (P_1, V_2) (b) an isochoric reduction in pressure to the state (P_2, V_2) (c) an isobaric reduction in volume to the state (P_2, V_1) (d) an isochoric increase in pressure back to the original state (P_1, V_1) again. What work is done in this cycle? If $P_1 = 3$ atm, $P_2 = 1$ atm, $V_1 = 1$ litre and $V_2 = 2$ litres, how much work is done *by* the gas in traversing the cycle 100 times?

3 During a reversible adiabatic expansion of an ideal gas, the pressure and volume at any moment are related by $PV^{\gamma} = c$ where c and γ are constants. Show that the work done *by* the gas in expanding from a state (P_1, V_1) to a state (P_2, V_2) is

$$W = \frac{P_1 V_1 - P_2 V_2}{\gamma - 1}$$

4 Ice at $0\,°C$ and at a pressure of 1 atm, has a density of 916.23 kg m^{-3}, while water under these conditions has a density 999.84 kg m^{-3}. How much work is done against the atmosphere when 10 kg of ice melt into water?

5 A metal container, of volume V and with diathermal walls, contains n moles of an ideal gas at high pressure. The gas is allowed to leak out slowly from the container through a small valve to the atmosphere at a pressure P_0. The process occurs isothermally at the temperature of the surroundings. Show that the work done by the gas against the surrounding atmosphere is

$$W = P_0(nv_0 - V)$$

where v_0 is the molar volume of the gas at atmospheric pressure and temperature.

6   Show that the coefficient of linear expansion α is related to the expansivity β as

$$\beta = 3\alpha$$

7   A hypothetical substance has an isothermal compressibility $\kappa = a/v$ and an expansivity $\beta = 2bT/v$ where a and b are constants and v is the molar volume. Show that the equation of state is

$$v - bT^2 + aP = \text{a constant}$$

8   A welded railway line, of length 15 km, is laid without expansion joints in a desert where the night and day temperatures differ by 50 K. The cross-sectional area of the rail is 3.6×10^{-3} m^2. (a) What is the difference in the night and day tension in the rail if it is kept at constant length? (b) If the rail is free to expand, by how much does its length change between night and day? ($\alpha = 8 \times 10^{-6}$ K^{-1}; $Y = 2 \times 10^{11}$ N m^{-2}.)

9   Find an expression for the work done when a wire of length L is heated reversibly from a temperature T_1 to a temperature T_2 under conditions of constant tension $\mathscr{F}$.

10 The equation of state of a rubber band is

$$\mathscr{F} = aT\left[\frac{L}{L_0} - \left(\frac{L_0}{L}\right)^2\right]$$

where L_0 is the original length and a is a constant equal to 1.3×10^{-2} N K^{-1}. How much work is performed when the band is stretched isothermally and reversibly from its original length of 10 cm to 20 cm, the temperature being 20 °C?

11 A block of metal at a pressure of one atmosphere is initially at a temperature of 20 °C. It is heated reversibly to 32 °C at constant volume. Calculate the final pressure. If the heating had been carried out irreversibly, would this affect your answer? (The expansivity $\beta = 5.0 \times 10^{-5}$ K^{-1}; the isothermal bulk modulus $K = 1.5 \times 10^{11}$ N m^{-2}.)

Chapter 3

1   Liquid is stirred at constant volume inside a container with adiabatic walls. The liquid and the container are regarded as the system. (a) Is heat being transferred to the system? (b) Is work being done on the system? (c) What is the sign of the internal energy change of the system?

2   Water inside a rigid cylindrical insulated tank is set into rotation and left to come to rest under the action of viscous forces. Regard the tank and the water as the system. (a) Is any work done by the system as the water comes to rest? (b) Is there any heat flow to or from the system? (c) Is there any change in the internal energy of the system?

3   A combustion experiment is performed on a mixture of fuel and oxygen contained in a constant-volume container surrounded by a water bath. The temperature of the water is observed to rise. Regard the matter inside the container as the system. (a) Has work been done on the system? (b) Has heat been transferred between the system and the surroundings? (c) What is the sign of the internal energy change of the system?

4   A gas is contained in a cylinder fitted with a frictionless piston and is taken from the state a to the state b along the path acb shown in Figure A4.1. 80 J of heat flow into the system and the system does 30 J of work. (a) How much heat flows into the system along the path adb if the work done by the gas system is 10J? (b) When the system is returned from b to a along the curved path, the work done on the system is 20 J. What is the heat transfer? (c) If $U_a = 0$ and $U_d = 40$ J, find the heat absorbed in the processes ad and db.

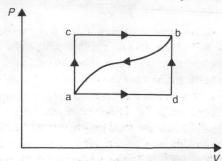

Fig. A4.1

5   An electrical resistance coil, wired to the surroundings, is placed inside a cylinder fitted with a frictionless piston and containing an ideal gas. The walls of the cylinder and the piston are adiabatic. A current of 5 A is maintained through the resistance, across which there is a voltage drop of 100 V. The piston is opposed by a constant external force of 5000 N. At what speed must the piston move outwards in order that there is no change in the temperature of the gas? Is the electrical energy transferred to the gas as heat or work?

Suppose now that the walls are diathermal and the resistance coil is wrapped round the outside of the cylinder. Regard the system as the cylinder and the gas, excluding the heating coil. Is the energy transfer now heat or work?

6   Two moles of a monatomic ideal gas are at a temperature of 300 K. The gas expands reversibly and isothermally to twice its original volume. Calculate the work done by the gas, the heat supplied and the change in the internal energy.

7   An ideal gas is contained in a cylinder fitted with a frictionless piston at the pressure P and volume V. It is heated quasistatically and at constant volume so that its temperature is doubled, and then it is cooled at constant pressure until it returns to its original temperature. Show that the work done on the gas is PV.

8   Find the change in the internal energy of one mole of a monatomic ideal gas in an isobaric expansion at 1 atm from a volume of 5 m^3 to a volume of 10 m^3. γ for a monatomic ideal gas is 5/3.

9   The molar specific heat of many materials at low temperatures is found to obey the Debye law $c_v = A\,[T/\theta]^3$ where A is a constant equal to 1.94×10^3 J mol^{-1}K^{-1} and with the Debye temperature θ taking different values for different materials. For diamond, it is 1860 K. (a) Evaluate c_v at 20 K and 100 K. (b) How much heat is required to heat one mole of diamond between 20 K and 100 K? (c) What is the average molar specific heat in this range?

10 Show that the adiabatic curve for an ideal gas is steeper by a factor of γ than the isotherm through a point on the P-V indicator diagram.

11 Show that the following relations hold for a reversible adiabatic expansion of an ideal gas:

$$TV^{\gamma-1} = \text{a constant}$$

$$\frac{T}{P^{1-1/\gamma}} = \text{another constant}$$

The fireball of a uranium fission bomb consists of a sphere of gas of radius 15 m and temperature 300 000 K shortly after detonation. Assuming that the expansion is adiabatic and that the fireball remains spherical, estimate the radius of the ball when the temperature is 3000 K. (Take $\gamma = 1.4$.)

12 An interstellar cloud, made up of an ideal gas, collapses with its radius decreasing as

$$R = 10^{13}\left(\frac{-t}{216}\right)^{\frac{2}{3}} \text{m}$$

with t measured in years. The time t is taken to be zero at zero radius so that t is always negative. The cloud collapses isothermally at 10 K until its radius reaches 10^{13} m. It then becomes opaque so that, from then on, the collapse takes place adiabatically ($\gamma = 5/3$) and reversibly. How many years does it take for the temperature to rise by 800 K measured from the time the cloud reaches a radius of 10^{13} m?

13 A thick-walled insulating chamber contains n_1 moles of helium gas at a high pressure P_1 and temperature T_1. It is allowed to leak out slowly to the atmosphere at a pressure P_0 through a small valve. Show that the final temperature of the n_2 moles of helium left in the chamber is

$$T_2 = T_1\left(\frac{P_0}{P_1}\right)^{\frac{\gamma-1}{\gamma}} \qquad \text{with} \qquad n_2 = n_1\left(\frac{P_0}{P_1}\right)^{1/\gamma}$$

(Hint: Consider as your system the gas that is ultimately left in the chamber.)

14 Calculate the work done by a van der Waals gas, with equation [3.14] as the equation of state, in expanding from a volume V_1 to a volume V_2: (a) at constant pressure P (b) at constant temperature T.

15 A magnetic salt obeys the Curie law

$$\frac{\mu_0 M}{B_0} = \frac{\mathscr{C}}{T}$$

where M is the magnetisation, B_0 is the applied induction in the absence of the specimen, $\mathscr{C}$ is a constant and μ_0 is the permeability of free space. The salt is magnetised isothermally from a magnetisation M_1 to M_2. You may assume that the magnetisation is uniform over the volume of the salt. Show that the work of magnetisation is

$$W = \frac{V\mu_0 T}{2\mathscr{C}} \left(M_2{}^2 - M_1{}^2\right)$$

16 The infinitesimal work done in charging a cell is $dW = \mathscr{E}\, dZ$ so the rate of doing work is $\mathscr{E}\, dZ/dt = \mathscr{E}I$ where I is the current supplied. A battery is charged by applying a current of 40 A at 12 V for 30 minutes. During this charging process the battery loses 200 kJ of heat to the surroundings. By how much does the internal energy of the cell change, assuming that there are no forms of work other than electrical work?

17 A steam turbine takes in steam at the rate of 6000 kg hour^{-1} and its power output is 800 kW. Neglect any heat loss from the turbine. Find the change in the specific enthalpy of the steam as it passes through the turbine if (a) the entrance and exit are at the same elevation and the entrance and exit velocities are negligible (b) the entrance velocity is 50 m s^{-1} and the exit velocity is 200 m s^{-1}, with the outlet pipe being 2 m above the inlet.

Chapter 4

1 Heat is supplied to an engine at the rate of 10^6 J minute^{-1} and the engine has an output of 10 horsepower. What is the efficiency of the engine and what is the heat output per minute?

2 A storage battery delivers a current into an external circuit and performs electrical work. The battery remains at a constant temperature by absorbing heat from the surrounding atmosphere. Heat then appears to be completely converted into work. Is this a violation of the second law?

3 Show that two adiabatics cannot intersect. (Hint: Imagine that they do; complete a cycle with an isotherm and operate an engine round this cycle.)

4 An inventor claims to have developed an engine which takes in 11×10^7 J at 400 K, rejects 5×10^7 J at 200 K and delivers 16.67 kW hours of work. Would you advise investing money in this project?

5 Which gives the greater increase in the efficiency of a Carnot engine: increasing the temperature of the hot reservoir or lowering the temperature of the cold reservoir by the same amount?

6 50 kg of water at 0 °C has to be frozen into ice in a refrigerator. The room temperature is 20 °C. What is the minimum work input to the refrigerator to achieve this. (Latent heat of fusion of water $= 3.33 \times 10^5$ J kg^{-1}.)

7 It is proposed to heat a house using a heat pump operating between the house and the outside. The house is to be kept at 22 °C, the outside is at -10 °C and the heat loss from the house is 15 kW. What is the minimum power required to operate the pump?

8

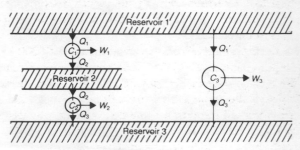

Fig. A4.2

Show that the efficiencies of the three Carnot engines, operating between the three reservoirs as illustrated in Fig. A4.2, are related as

$$\eta_3 = \eta_1 + \eta_2 - \eta_1\eta_2$$

9 A Carnot engine working on a satellite in outer space has to deliver a fixed amount of power at the rate $\dot{W}$. The temperature of the heat source is also fixed, at T_1. The lower temperature reservoir at T_2 consists of a large body of area A; its temperature is maintained at T_2 because it radiates energy into space as much heat as is delivered to it by the engine. The rate of this radiation is $\sigma A T_2^4$ where σ is a constant. The Carnot engine has to be designed so that, for a given $\dot{W}$ and T_1, A has a minimum value. Show that A has a minimum value when T_2 takes the value $3T_1/4$.

10

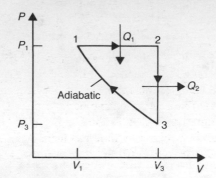

Fig. A4.3

A hypothetical engine, with an ideal gas as the working substance, operates in the cycle shown in Fig. A4.3. Show that the efficiency of the engine is

$$\eta = 1 - \frac{1}{\gamma}\left(\frac{1 - P_3/P_1}{1 - V_1/V_3}\right)$$

11

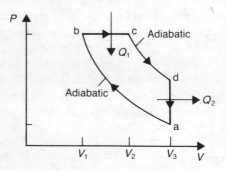

Fig. A4.4

A simplified representation of the diesel cycle, with just air as the working substance, is as shown in Fig. A4.4. Show that the efficiency of this engine is

$$\eta = 1 - \frac{1}{\gamma}\left(\frac{\frac{1}{r_e{}^{\gamma}} - \frac{1}{r_c{}^{\gamma}}}{\frac{1}{r_e} - \frac{1}{r_c}}\right)$$

where $r_e = V_3/V_2$, the expansion ratio, and $r_c = V_3/V_1$, the compression ratio. If $r_e = 5, r_c = 15$ and $\gamma = 1.4$, evaluate η. Notice that the compression ratio can be very much higher in a diesel engine than in a petrol engine because diesels do not suffer from pre-ignition, or pinking, as the fuel is sprayed in at the end of the compression stroke; this allows an increased r_c. This is one reason why diesels are more efficient than petrol engines.

Chapter 5

1 A bucket containing 5 kg of water at 25 °C is put outside a house so that it cools to the temperature of the outside at 5 °C. What is the entropy change of the water? (c_P for water = 4.19×10^3 J kg^{-1})

2 Calculate the entropy change for each of the following: (a) 10 g of steam at 100 °C and a pressure of one atmosphere condensing into water at the same temperature and pressure. (The latent heat of vaporisation of water is 2257 J g^{-1}.) (b) 10 g of water at 100 °C and a pressure of one atmosphere cooling to 0 °C at the same pressure. (The average specific heat of water between 0 °C and 100 °C is 4.19 J g^{-1}.) (c) 10 g of water at 0 °C and a pressure of one atmosphere freezing into ice at the same pressure and temperature. (The latent heat of fusion of ice is 333 J g^{-1}.)

3 The low temperature molar specific heat of diamond varies with temperature as:

$$c_v = 1.94 \times 10^3 \left[\frac{T}{\theta}\right]^3 \qquad \text{J mol}^{-1}\,\text{K}^{-1}$$

where the Debye temperature θ = 1860 K. What is the entropy change of 1 g of diamond when it is heated at constant volume from 4 K to 300 K? (The atomic weight of carbon is 12.)

4 An electric current of 10 A flows for one minute through a resistor

of 20 ohms which is kept at 10 °C by being immersed in running water. What is the entropy change of the resistor, the water and the universe?

5 A thermally insulated resistor of 20 ohms has a current of 5 A passed through it for 1 s. It is initially at 20 °C. (a) What is the temperature rise? (b) What is the entropy change of the resistor and the universe? Mass of resistor is 5 g; c_P for the resistor is 0.8×10^3 J kg^{-1} K^{-1}. (Hint: In the actual process, dissipative work is done on the resistor. Imagine a reversible process taking it between the same equilibrium states.)

6 An ideal gas has a molar specific heat given by $c_v = A + BT$ where A and B are constants. Show that the change in entropy per mole in going from the state $(V_1 T_1)$ to the state $(V_2 T_2)$ is

$$\Delta s = A \ln (T_2/T_1) + B (T_2 - T_1) + R \ln (V_2/V_1)$$

7 A 50 kg bag of sand at 25 °C falls 10 m onto the pavement and comes to an abrupt stop. What is the entropy increase of the sand? Neglect any transfer of heat between the sand and the surroundings and assume that the thermal capacity of the sand is so large that its temperature is unchanged. (Hints. Ask yourself: (a) What is the dissipative work done on the sand? (b) What is the change in the internal energy of the sand? (c) What is the entropy change associated with this ΔU at constant T? The sand does no work as it deforms when it hits the pavement: only its shape changes, not its volume.)

8 One mole of an ideal gas undergoes a free expansion tripling its volume. What is the entropy change of (a) the gas (b) the universe?

9 Two equal quantities of water, of mass m and at temperatures T_1 and T_2, are adiabatically mixed together, the pressure remaining constant. Show that the entropy change of the universe is

$$\Delta S = 2mc_P \ln \left\{ \frac{T_1 + T_2}{2 \sqrt{T_1 T_2}} \right\}$$

where c_P is the specific heat of water at constant pressure. Show that $\Delta S > 0$. (Hint: $(a - b)^2 > 0$ for a and b real.)

10 Consider two identical bodies of heat capacity C_P and with negligible thermal expansion coefficients. Show that when they are placed in thermal contact in an adiabatic enclosure their final

temperature is $(T_1 + T_2) / 2$ where T_1 and T_2 are their initial temperatures.

Now consider these two bodies being brought to thermal equilibrium by a Carnot engine operating between them. The size of the cycle is small so that the temperatures of the bodies do not change appreciably during one cycle; thus the bodies behave as reservoirs during one cycle. Show that the final temperature is $(T_1 T_2)^{1/2}$. (Hint: What is the entropy change of the universe for this second process?)

11   Let us define a semipermeable membrane as one which allows the passage of one type of molecule. At equilibrium the gas pressures on either side of such a membrane are equal. Such membranes exist.

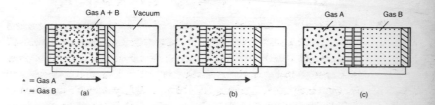

Fig. A4.5

Consider a mixture of two ideal gases A and B contained in the left-hand half of the box as shown in Fig. A4.5(a). There is a vacuum in the right-hand half. The box is fitted with a pair of coupled sliding pistons; the left-hand one is permeable to A only, while the right-hand one is impermeable to both. The box is divided into two with a partition permeable to B only. Now slide the coupled pistons slowly to the right so that, eventually, the two gases separate reversibly. They will finally each occupy a volume equal to the original volume of the mixture. This is shown in Fig. 4.5(c). Let this process occur isothermally. (a) By considering the pressures due to each gas on either side of the membranes, show that the net force on the coupled pistons is zero. (b) The heat flowing into the system in this isothermal reversible process is $Q = T (S_f - S_i)$ where S_i and S_f are the initial and final entropies. (c) By now applying the first law, show that $S_i = S_f$.

This result is known as Gibbs's theorem. It is saying that:

In a mixture of ideal gases the entropy is the sum of the entropies that each gas would have if it alone occupied the whole volume, i.e.

$$S_{A+B}(T, V) = S_A(T, V) + S_B(T, V)$$

12 We have seen that for n moles of an ideal gas the entropy is

$$S = nC_V \ln T + n R \ln (V/n) + S_0 \qquad [5.11]$$

n_A moles of an ideal gas A of volume V_A and temperature T are separated from n_B moles of another ideal gas B of volume V_B at the same temperature T (see Fig. A4.6(a)). The partition is removed so that the gases mix isothermally at the temperature T, the mixture then occupying the volume $V_A + V_B$ (see Fig. A4.6(b)).

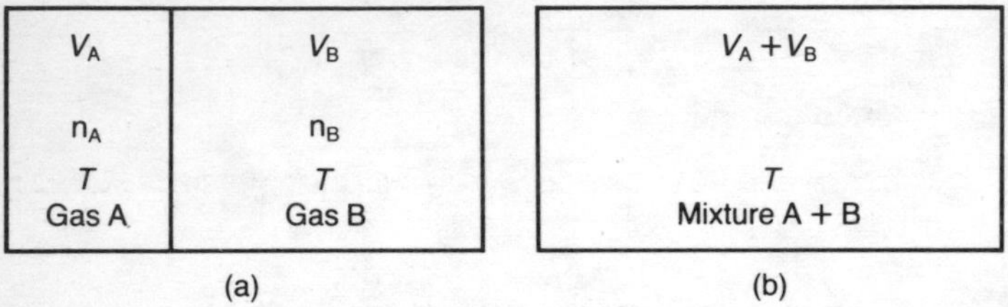

V_A	V_B		$V_A + V_B$
n_A	n_B		
T	T		T
Gas A	Gas B		Mixture A + B
(a)			(b)

Fig. A4.6

(a) Use Gibbs's theorem, introduced in the previous question, to show that the entropy change occurring in this mixing is

$$\Delta S_{mixing} = R \left[n_A \ln \left\{ \frac{V_A + V_B}{V_A} \right\} + n_B \ln \left\{ \frac{V_A + V_B}{V_B} \right\} \right]$$

(b) Suppose that the gases are identical. Clearly, on removing the partition, there can now be no entropy change as the physical system is unchanged, yet the result you have just proved in (a) gives $\Delta S_{mixing} \neq 0$! This is known as the Gibbs paradox. Is the result given in (a) valid for identical gases and if not, why not? (Hint: Consider how Gibbs's theorem was proved.) (c) Obtain the correct expression

$$\Delta S_{mixing} = (n_A + n_B)R \ln \left(\frac{V_A + V_B}{n_A + n_B} \right)$$

$$- n_A R \ln \frac{V_A}{n_A} - n_B R \ln \frac{V_B}{n_B}$$

for the entropy of mixing of identical gases by applying equation [5.11] to the three volumes V_A, V_B and $V_A + V_B$, all containing the same gas. (d) By using the fact that, for identical gases,

$$\frac{V_A + V_B}{n_A + n_B} = \frac{V_A}{n_A} = \frac{V_B}{n_B}$$

show that the entropy of mixing given in (c) is indeed zero. (Gibbs's paradox is discussed nicely in the little book by Chambadal, see Appendix 6.)

Chapter 6

1 It is a result of statistical mechanics that the internal energy of an ideal gas is

$$U = U(S, V) = \alpha N k_B \left(\frac{N}{V} \right)^{2/3} e^{\frac{2S}{3Nk_B}}$$

where α is a constant and the other symbols have their usual meanings. Show that the equation of state $PV = nRT$ follows from this equation.

2 The Helmholtz function of one mole of a certain gas is:

$$f = F/n = -a/v - RT \ln (v - b) + j(T)$$

where a and b are constants and j is a function of T only. Derive an expression for the pressure of the gas.

3 The table gives the values of some thermodynamic properties of a substance at two different states, both at the same temperature.

Temp.	u	s	P	Specific volume
°C	kJ kg^{-1}	kJ K^{-1}kg^{-1}	N m^{-2}	m^3 kg^{-1}
Initial state 300	2727	6.364	4 × 10^6	0.0588
Final state 300	2816	8.538	0.05 × 10^6	5.29

What is the maximum amount of work that can be extracted from one kilogram of this substance in taking it from the initial state to the final state? You will have to select the relevant data from the table. This substance, incidentally, is superheated steam.

4   The Gibbs function of one mole of a certain gas is given by

$$g = RT \ln P + A + BP + CP^2/2 + DP^3/3$$

where A, B, C and D are constants. Find the equation of state of the gas.

5   Derive the following equations:

(a) $U = F - T\left(\dfrac{\partial F}{\partial T}\right)_V = -T^2\left(\dfrac{\partial F/T}{\partial T}\right)_V$

(b) $C_V = -T\left(\dfrac{\partial^2 F}{\partial T^2}\right)_V$

(c) $H = G - T\left(\dfrac{\partial G}{\partial T}\right)_P = -T^2\left(\dfrac{\partial G/T}{\partial T}\right)_P$

(d) $C_P = -T\left(\dfrac{\partial^2 G}{\partial T^2}\right)_P$

6   In the presence of a catalyst, one mole of NO decomposes into nitrogen and oxygen. The initial and final temperatures are 25° C and the process occurs at a pressure of one atmosphere. The entropy change is $\Delta s = 76$ J K^{-1} mol^{-1} and the enthalpy change is $\Delta H = -8.2 \times 10^5$ J mol^{-1}. What is the change in the Gibbs free energy and what is the heat produced in the decomposition?

7   A gas cools from a temperature T to the temperature T_0 of the surroundings. There is no change between the initial and final volumes, $\Delta V = 0$, but the volume may *vary* during the process and so the gas may perform work. This is indicated in Fig. A4.7.

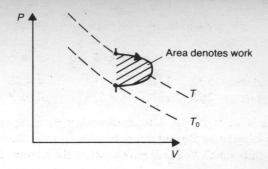

Fig. A4.7

Show that the maximum amount of work obtainable from the gas is

$$W_{max} = C_V(T - T_0) + C_V T_0 \ln T_0/T$$

(Hints: (a) Consider the argument leading to equation [6.28] (b) Hence show $W_{max} \leqslant -\Delta U + T_0 \Delta S$ (c) Use equation [5.11]

8 One mole of an ideal gas expands at the constant temperature T_0 of the surroundings from a pressure P_1 to a pressure P_2. The atmospheric pressure is P_0. (a) By considering the total work done in a reversible expansion and subtracting off the useless work, show that the maximum useful work done by the gas is

$$RT_0 \ln\left(\frac{P_1}{P_2}\right) - P_0 RT_0 \ (1/P_2 - 1/P_1)$$

(b) How is this work related to the change in the Gibbs function?

9 In a simple form of fuel cell, which is a device for producing electricity from a chemical reaction, hydrogen gas is fed in at one electrode and oxygen at the other. Water is produced according to the reaction

$$2H_2 + O_2 \rightarrow 2H_2O$$

The cell operates at the pressure and temperature (298 K) of the atmosphere. Assuming that the cell operates reversibly, calculate its EMF, given the following molar values for S and H:

	s	h
	$(J\,K^{-1}\,mol^{-1})$	$(J\,mol^{-1})$
O_2	201	17.2×10^3
H_2	128	8.1×10^3
H_2O	66.7	-269×10^3

(Hint: Consider one mole of H_2O being produced. The useful work done by the cell is the $2\,F_0\,\mathcal{E}$. For this isobaric isothermal process, we may use equation [6.49].)

Chapter 7

1   Derive the relation

$$\left(\frac{\partial C_P}{\partial P}\right)_T = -T\left(\frac{\partial^2 V}{\partial T^2}\right)_P$$

2   Show that C_V for a van der Waals gas is a function of temperature only.

3   Derive the second energy equation

$$\left(\frac{\partial U}{\partial P}\right)_T = -\left\{T\left(\frac{\partial V}{\partial T}\right)_P + P\left(\frac{\partial V}{\partial P}\right)_T\right\}$$

4   Consider n moles of a van der Waals gas. Show that $\left(\frac{\partial U}{\partial V}\right)_T = \frac{n^2 a}{V^2}$. Hence show that the internal energy is

$$U = \int_0^T C_V dT - a\,n^2/V + U_0$$

where U_0 is a constant. {Hint: Express $U = U(T, V)$}.

5   As in the previous question, consider n moles of a van der Waals gas. Show that

(a) $$S = \int_0^T \frac{C_V}{T}\,dT + nR\ln(V - nb) + S_0$$

where S_0 is a constant. {Hint: Use $dS = 1/T\,(dU + PdV)$}.
(b) The equation for a reversible adiabatic process is

$$T(V - n\,b)^{nR/C_V} = \text{a constant}$$

if C_V is assumed to be independent of T.

6 Show that the difference between the isothermal and the adiabatic compressibilities is

$$\kappa_T - \kappa_S = T\frac{V\beta^2}{C_P}$$

7 For each of the following processes, state whether the process is reversible or irreversible and state which of the quantities S, H, U, F and G are unchanged: (a) An isothermal quasistatic expansion of an ideal gas in a cylinder fitted with a frictionless piston. (b) As (a), but for a non-ideal gas. (c) A quasistatic adiabatic expansion of a gas in a cylinder fitted with a frictionless piston. (d) An adiabatic expansion of an ideal gas into a vacuum (a free expansion). (e) A throttling process of a gas through a porous plug (the Joule–Kelvin effect).

8 We have seen that a Carnot cycle takes a particularly simple rectangular form on an S–T plot (see Fig. 5.10). An S–H plot is also useful in engineering. Show that, for an ideal gas as the working substance, a Carnot cycle is again rectangular in this representation. Hint. First derive the general thermodynamic relations:

$$\left(\frac{\partial H}{\partial S}\right)_T = T - V\left(\frac{\partial T}{\partial V}\right)_P$$

$$\left(\frac{\partial H}{\partial T}\right)_S = C_P\frac{V}{T}\left(\frac{\partial T}{\partial V}\right)_P$$

9 Derive the so-called $T\mathrm{d}S$ equations:

$$T\mathrm{d}S = C_V\mathrm{d}T + T\left(\frac{\partial P}{\partial T}\right)_V \mathrm{d}V$$

$$T\mathrm{d}S = C_P\mathrm{d}T - T\left(\frac{\partial V}{\partial T}\right)_P \mathrm{d}P$$

$$T\mathrm{d}S = C_V\left(\frac{\partial T}{\partial P}\right)_V \mathrm{d}P + C_P\left(\frac{\partial T}{\partial V}\right)_P \mathrm{d}V$$

10 A block of metal of volume V is subjected to an isothermal reversible increase in pressure from P_1 to P_2 at the temperature T. (a) Show that the heat given out by the metal is $TV\beta(P_2 - P_1)$.

(b) Show that the work done on the metal is $V(P_2{}^2 - P_1{}^2)/2K$.

(c) By using the first law, calculate the change in U. (d) Obtain the same result as in (c) by writing $U = U(T,P)$ so

$$dU = \left(\frac{\partial U}{\partial T}\right)_P dT + \left(\frac{\partial U}{\partial P}\right)_T dP$$

You may assume that β, K and V are approximately constant during the compression.

11 A block of metal is subjected to an adiabatic and reversible increase of pressure from P_1 to P_2. Show that the initial and final temperatures T_1 and T_2 are related as

$$\ln T_2/T_1 = \frac{V\beta}{C_P}(P_2 - P_1)$$

You may assume that the volume of the block stays constant during the compression.

12 Assuming that helium obeys the van der Waals equation of state, determine the change in temperature when one kilomole of helium gas, initially at $20\,°C$ and with a volume of $0.12\ m^3$, undergoes a free expansion to a final pressure of one atmosphere. You should first show that

$$\left(\frac{\partial T}{\partial V}\right)_U = -\frac{a}{C_V}\left(\frac{n}{V}\right)^2$$

$a = 3.44 \times 10^3\ J\,m^3\ kmol^{-2}$; $b = 0.0234\ m^3\ kmol^{-1}$; $c_V/R = 1.506$.
(Hint: You may approximate. First show that $P_1 \gg P_2$. Then you may take $V_2 \gg V_1$.)

13 Show that the Joule–Kelvin coefficient is zero for an ideal gas.

14 One kilomole of an ideal gas undergoes a throttling process from $P_1 = 4$ atm to $P_2 = 1$ atm. The initial temperature of the gas is $50\,°C$. (a) What is the temperature change? (b) How much work would have to be done on the gas to take it reversibly between the initial and final states? (c) What is the entropy change of the gas? (Hint: You should calculate the entropy change in two ways: (i) Imagine a reversible process in which the gas is taken isothermally between the initial and final states. (ii) Apply the general approach of writing $S = S(T, H)$ and then imagining a reversible process in which the pressure is changed at constant H. You will

need to show that $(\partial S/\partial P)_H = -V/T$ in general, from $dH = T\,dS + V\,dP$, and this is $-nR/P$ for an ideal gas.)

15   Show that the temperature and volume of the points (T_{in}, V_{in}) on the inversion curve for a van der Waals gas undergoing a Joule–Thompson expansion are related as

$$T_{in} = 2a\,(V_{in} - nb)^2\,(Rb\,V_{in}{}^2)^{-1}$$

Assuming that, at the maximum inversion temperature, $V_{in} \gg nb$, show that $T_{in}{}^{max} \approx 2a/Rb$.

Chapter 8

1   A paramagnetic salt is magnetised isothermally and reversibly from zero applied induction to a final induction field of B_0. It obeys the Curie law $\chi_m = \mathscr{C}/T$. Show that the heat of magnetisation is

$$Q = -\frac{\mathscr{C}V}{T\mu_0}\frac{B_0{}^2}{2}$$

where V is the volume of the salt. Is this heat absorbed or rejected? (Hint: Express $S = S\,(T, B_0)$.)

2   The paramagnetic salt of the previous question is adiabatically and reversibly demagnetised from an initial state (T_1, B_{01}) to a final state $(T_2, 0)$. The internal energy $U = \alpha T^4$ where α is a constant. Show that

$$T_2^3 = T_1^3 - \frac{3\mathscr{C}VB_{01}^2}{8\alpha\mu_0 T_1^2}$$

(Hint: The natural way to tackle this problem is to write $T = T\,(S, B_0)$. While the problem can be solved in this way, it then involves a *very* nasty integral. Instead write $T = T\,(S,\,\mathscr{M})$ and recognise that $\mathscr{M}$ decreases to zero at the end of the demagnetisation.)

3   Calculate the temperature of the sun, assuming it to be a black body, if the solar energy falling on the surface of the earth is 1.4×10^3 J m^{-2} s^{-1} (this is the solar constant). The diameter of the sun is 13.9×10^8 m and the mean distance of the sun from the earth is 14.9×10^{10} m. (You are probably surprised by the value of the solar constant; this is no less than 1 kW of power for each 0.7 m^2 of the earth's surface! This explains the interest in solar panels.)

4 Thermal radiation is enclosed in a cavity of volume V. Show that the entropy associated with this radiation is

$$S = \frac{16\,\sigma\,T^3\,V}{3c} + \text{a constant}$$

Hence, or otherwise, show that if the radiation field is expanded isentropically,

$$P V^{\frac{4}{3}} = \text{a constant}$$

where P is the radiation pressure and the other symbols have their usual meanings. This result is important in astrophysics. (Hint: Use $dS = (dU + P\,dV)/T$.)

5 The infinitesimal work term for a stretched wire is $dW = \mathscr{F}\,dL$. By considering the four potential functions U, H, F and G for this system, derive the four Maxwell relations and show that they are consistent with the mnemonic

$$S$$

$$-\mathscr{F} \qquad L$$

$$T$$

6 The equation of state of a rubber band is

$$\mathscr{F} = a\,T\left[\frac{L}{L_0} - \left(\frac{L_0}{L}\right)^2\right]$$

where $\mathscr{F}$ is the tension, L the length, L_0 the unstretched length of 1 m and $a = 1.3 \times 10^{-2}$ N K^{-1} is a constant. (a) Derive the energy equation for the band

$$\left(\frac{\partial U}{\partial L}\right)_T = \mathscr{F} - T\left(\frac{\partial \mathscr{F}}{\partial T}\right)_L$$

(b) Show that U is a function of T only. (c) Find the work done on the band and the heat rejected when it is stretched isothermally and reversibly at 300 K from 1 m to 2 m. (d) If the band is stretched isentropically from 1 m to 2 m, what is the final temperature? ($C_L = 1.2$ J K^{-1}.)

7 Consider an electrolytic cell operating under reversible, isothermal, isobaric conditions and with no change in volume, as discussed in section 8.11. Also, consider a reaction involving standard molar quantities of the electrolytes. (a) Show that the changes in the entropy, the Helmholtz and the Gibbs functions are

$$\Delta S = n F_0 \frac{d\mathscr{E}}{dT}$$

$$\Delta F = \Delta G = -n F_0 \mathscr{E}$$

(b) The EMF varies with temperature as in the following table:

EMF (mV)	T(K)
213	1054
212	1086
209	1106
202	1174
200	1214
198	1244
194	1283

What is the entropy change in this range of temperatures? What is the corresponding change in G at 1100 K? The number of electrons transferred per ion is 2.

Chapter 9

1

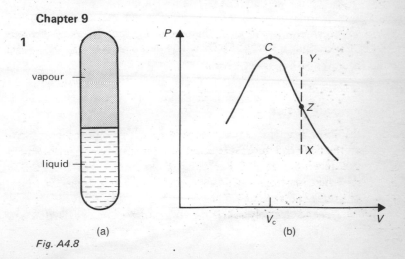

Fig. A4.8

A small amount of pure liquid is contained in a glass tube from which all the air has been removed (Fig. A4.8(a)). The volume of the tube is significantly greater than the critical volume V_c of the enclosed liquid. (a) Describe what happens as the temperature is raised so that the substance traverses the path XY on the P-V projection in Fig. A4.8(b). (b) If the volume of the tube were equal to V_c, what would you observe as the temperature is raised?

2

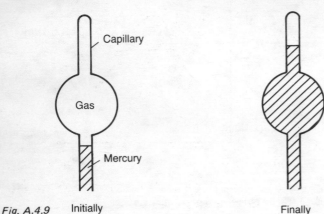

Fig. A.4.9   Initially   Finally

Gas is contained in a glass bulb of volume 250 cm³ (see Fig. A4.9). A capillary of length 10 cm and of diameter 1 mm is connected to the bulb. Mercury is forced into the bulb compressing the gas and forcing it into the capillary so that it occupies a length of 1 cm. This process occurs isothermally at 20 °C. The initial pressure of the gas is 10^{-3} torr. (a) What is the final pressure of the gas in the capillary if it is nitrogen? (b) What is the final pressure of the gas in the capillary if it is water vapour? (c) How much water condenses? Justify any assumptions. (The vapour pressure of water at 20 °C is 17.5 torr.)

3 Consider the isotherm at T on the P-V projection (Fig. A.4.10). At the point K the substance is a mixture of liquid and vapour. Let the masses of liquid and vapour be m_l and m_v and the total mass of the substance be m. Then the volume occupied by the mixture at K is $m_l v_l + m_v v_v$ where v_l and v_v are the specific volumes of the liquid and vapour. Let the specific volume of the mixture at K be v. Show that:

$$m_1(v - v_1) = m_v(v_v - v)$$

This result, which gives the ratio m_v/m_1, is known as the 'lever rule', for obvious reasons. m_v/m_1 is called the *quality* of the mixture.

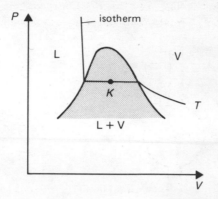

Fig. A4.10

4   At the critical point, $(\partial P/\partial V)_T = 0$ and $(\partial^2 P/\partial V^2)_T = 0$. Show that, for a van der Waals gas, the critical point is at:

$$P_c = \frac{a}{27b^2}, \quad V_c = 3nb, \quad T_c = \frac{8a}{27Rb}$$

5   The vapour pressure of camphor is as follows:

Temperature (°C)	30.8	55.0	62.0	78.0
Pressure (torr)	1.04	3.12	4.22	7.8

By plotting $\ln P$ against $1/T$, estimate the latent heat of vaporisation.

6   The equations of the sublimation and the vaporisation curves of a particular material are given by:

$$\ln P = 0.04 - 6/T \quad \text{sublimation}$$
$$\ln P = 0.03 - 4/T \quad \text{vaporisation}$$

where P is in atmospheres. (a) Find the temperature and pressure of the triple point. (b) Show that the molar latent heats of vaporisation and sublimation are $4R$ and $6R$. You may assume that the specific volume in the vapour phase is much larger than those in the liquid and solid phases. (c) Find the latent heat of fusion. (Hint: Consider a loop round the triple point in the P-T projection. As S is a state function,

$$\{l_{SV}/T_{TP}\} - \{l_{SL}/T_{TP}\} - \{l_{LV}/T_{TP}\} = 0.)$$

7

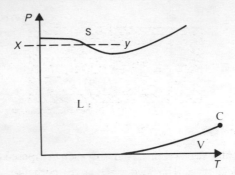

Fig. A4.11

The phase diagram for ^{3}He is as in Fig. A4.11. Discuss the variation of G along the section XY. What does this tell you about the entropy of the solid phase compared with the entropy of the liquid phase at the lowest temperatures? The result should surprise you.

8 Tin can exist in two forms, grey tin and white tin. Grey tin is the stable form at low temperatures and white tin the stable form at high temperatures. There is a first-order transition between the two phases with a transition temperature of 291 K at a pressure of 1 atm. What is the change in this transition temperature if the pressure is increased to 100 atm. (The latent heat for the transition is 2.20×10^3 J mol^{-1}; the densities of grey and white tin are 5.75×10^3 kg m^{-3} and 7.30×10^3 kg m^{-3}; and the atomic weight of tin is 119.)

9 The Curie temperature for nickel for the phase change from the ferromagnetic phase to the paramagnetic phase is 630 K, the pressure being 1 atmosphere. If the pressure is increased by 100 atmospheres, calculate the shift in the Curie temperature. (In this phase transition, c_P changes by 6.7 J K^{-1} mol^{-1} while β changes by 5.5×10^{-6} K^{-1}. The density of nickel is 8.9×10^3 kg m^{-3} and the atomic weight is 58.7.)

Chapter 10

1 Derive equation [10.3]. (Hint: Express $U = U(S, V, N_1, N_2 \ldots)$ and so d$U = \ldots$)

2

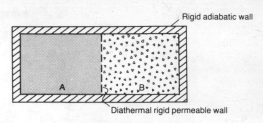

Fig. A4.12

Consider two systems, A and B, each composed of the same single *particle type*. The two systems are contained in a chamber surrounded by rigid adiabatic walls and they are separated from each other within the chamber by a rigid diathermal wall which is also permeable to the particles (see Fig. A4.12). Show, using an argument similar to the one used in section 10.2, that the condition for equilibrium against particle exchange is the equality of the chemical potentials.

3 Consider the system of question 2. Suppose that the two systems, composed of the same single type of particle, are *both in the same phase*, e.g. a gas on each side of the separating wall. Show that the pressures are equal. Would the pressures be equal if different phases existed on either side of the wall? (Hint: For a system consisting of a single type of particle, $\mu = G(T, P, N)/N = \phi(T, P)$. If the phases are the same on either side of the wall, the function ϕ must also be the same on either side.)

4 Consider the system of question 2 but suppose we now have a mixture of gases on either side of the separating wall which is permeable to all the different gases i. Argue that $\mu_A^i = \mu_B^i$ for all the different gases i. (Hint: Place another diathermal wall, permeable to only one gas, in front of the separating wall and proceed as in section 10.2.)

5 Show that the chemical potential of an ideal gas, at the temperature T, varies with pressure as

$$\mu = k_B T \ln (P/P_0) + \mu_0$$

where μ_0 is the value at reference point of pressure P_0 and temperature T. The gas consists of a single type of particle only. This expression is of great use in chemistry. (Hint: $V = (\partial G / \partial P)_{TN}$.)

6 Show that the chemical potential of an ideal monatomic gas of N particles is

$$\mu = \frac{5}{2} k_B T - \frac{3}{2} k_B T \ln T - k_B T \ln \left\{ \frac{N_A V}{N} \right\} - \frac{s_0 T}{N_A}$$

Hints:

 (i) Use equation [5.11] for S.

 (ii) $F = U - TS$ and $\mu = (\partial F / \partial N)_{TV}$

    or $G = U + PV - TS$ and $\mu = G/N$.

(iii) $U = 3/2 N k_B T$.

(iv) $C_V = 3/2 N k_B$.

Note: The N_A used here in this question is the Avogadro number and should not be confused with the notation N_A used in section 10.2 to denote the number of particles in a box A.

Appendix 5

Answers to questions

Chapter 1

1 327.79 degrees; 6.83 cm; no.
2 341.79 degrees.
3 348.35 K; 75.20 °C.
4 1.50×10^{-4} kg.
5 No, it is metastable; yes.

Chapter 2

1 5.7×10^4 J.
2 Gas does the amount of work $(P_1 - P_2)(V_2 - V_1)$; 2.02×10^4 J.
4 Atmosphere does 92.2 J of work on system.
8 (a) 2.88×10^5 N (b) 6 m.

9 $$W = \mathscr{F} \int_{T_1}^{T_2} \alpha L \, dT.$$

10 0.38 J.
11 892 atm; no difference.

Chapter 3

1 (a) No (b) Yes (c) ΔU positive and same as (b).
2 (a) No (b) No (c) It increases. Remember that, in $\Delta U = Q + W$, we exclude any changes in the bulk KE and PE (see p. 37). Here the total energy, U + the bulk (PE + KE), remains constant as there is no input of energy in the form of heat or work and the organised

motion of the rotational kinetic energy is converted into the random motion of internal energy (hence raising the temperature).

3   (a) No (b) Yes, Q negative (c) Negative.

4   (a) 60 J absorbed (b) 70 J rejected (c) 50 J and 10 J absorbed.

5   $10^{-1}\,ms^{-1}$; work; heat.

6   $3.46 \times 10^3\,J$; $3.46 \times 10^3\,J$; 0.

8   $7.6 \times 10^5\,J$.

9   (a) $2.41 \times 10^{-3}\,J\,K^{-1}\,mol^{-1}$; $3.02 \times 10^{-1}\,J\,K^{-1}\,mol^{-1}$ (b) 7.53 J mol^{-1} (c) $9.41 \times 10^{-2}\,J\,K^{-1}\,mol^{-1}$.

11 696 m.

12 208 years.

14 (a) $P(V_2 - V_1)$ (b) $nRT \ln\left\{\dfrac{V_2 - nb}{V_1 - nb}\right\} + n^2 a(1/V_2 - 1/V_1)$.

16 664 kJ.

17 (a) $-4.80 \times 10^5\,J\,kg^{-1}$ (b) $-4.99 \times 10^5\,J\,kg^{-1}$.

Chapter 4

1   44.8 per cent; $5.52 \times 10^5\,J$ minute^{-1}.

2   No, state of battery changed in process as Z changes.

4   No, $\eta = 54.6$ per cent while $\eta_C = 50$ per cent.

5   Lowering temperature of cold reservoir.

6   $1.22 \times 10^6\,J$.

7   1.63 kW.

11 56 per cent.

Chapter 5

1   $-1.46 \times 10^3\,J\,K^{-1}$.

2   (a) $-60.5\,J\,K^{-1}$ (b) $-13.1\,J\,K^{-1}$ (c) $-12.2\,J\,K^{-1}$.

3   $0.226\,J\,K^{-1}$.

4   0; $424\,J\,K^{-1}$; $424\,J\,K^{-1}$.

5   (a) 125 K (b) $1.42\,J\,K^{-1}$; $1.42\,J\,K^{-1}$.

7   $16.5\,J\,K^{-1}$.

8   (a) $9.13\,J\,K^{-1}$ (b) $9.13\,J\,K^{-1}$.

Chapter 6

2 $P = RT/(v - b) - a/v^2$. This is the van der Waals equation.

3 1.16×10^6 J.

4 $Pv = RT + BP + CP^2 + DP^3$ for one mole.

6 $\Delta g = -8.43 \times 10^5$ J mol^{-1}; 8.2×10^5 J mol^{-1} given out (exothermic).

8 (b) It is not related! Equation [6.49] gives $W^{useful} = -\Delta G$ for those processes where the end points are at (P_0, T_0). This is not so for this problem.

9 1.23 V.

Chapter 7

7 (a) Reversible; U, H (b) Reversible; none (c) Reversible; S (d) Irreversible; U, H (e) Irreversible; H.

10 (c) $- V\beta T(P_2 - P_1) + V(P_2^2 - P_1^2)/2 K$.

12 -2.3 K.

14 (a) 0 (b) 3.72×10^6 J (c) 1.15×10^4 J K^{-1}.

Chapter 8

3 5804 K.

6 (c) Work = 3.9 J; Heat rejected is 3.9 J (d) 303.3 K.

7 (b) $\Delta s = -16.4$ J K^{-1} mol^{-1}; $\Delta g = -4.04 \times 10^4$ J mol^{-1}.

Chapter 9

1 (a) The liquid level gradually goes down until only vapour is left at Z. (b) The liquid-vapour interface becomes blurred as the critical point is approached, with the vapour and the liquid becoming indistinguishable.

2 (a) 31.9 torr (b) 17.5 torr (c) 1.1×10^{-7} g. Neglect volume of water in capillary.

5 3.8×10^4 J mol^{-1}.

6 (a) 200 K, 1.01 atm (c) $2R$ mol^{-1}.

7 $S_{solid} > S_{liquid}$. You should discuss this with your quantum mechanics tutor.

8 -5.9 K.

9 0.03 K.

Appendix 6
Bibliography

Adkins, C.J. *Equilibrium Thermodynamics*, Cambridge University Press.

Chambadal, P. *Paradoxes in Physics*, Transworld.

Crangle, J. *The Magnetic Properties of Solids*, Arnold.

Dobbs, E.R. (1984) *Electricity and Magnetism*, Routledge & Kegan Paul.

Kaye, G. and Laby, T. *Tables of Physical and Chemical Constants* Longman.

Warn, J.R. *Concise Chemical Thermodynamics*, Van Nostrand.

Zemansky, M.W. and Dittman, R.H. *Heat and Thermodynamics* McGraw-Hill.

Index